The Landscapes of Italian Food

This book examines contemporary food systems in Italy, paying particular attention to the landscape, innovative local practices and local cultural history. It illustrates the utility of the value chain concept in navigating the complexities of comparative advantage in an advanced market setting. It establishes the connection between the landscape and individual food practices, and how they have responded to the commodification of the agri-food system, maintaining a distinctive local character while ensuring development and a healthy diet. It explores how community gardens are now a consolidated part of Italian urban experience, as well as the multiple policy frameworks which govern these activities. The book then explores a wider range of food procurement channels, from food cooperatives to buying groups and institutional partnerships, including the strategies employed by large retail groups to respond to the growing environmental sensitivity of their customers. Multifunctional implications of antimafia activities involving social agriculture are also explored. Finally, the book ends with a survey of European and domestic Italian policies aiming to protect and promote healthy food practices while preserving the integrity of the landscape.

This is fascinating reading for anyone interested in quality food and the territory, as well as academic readers from such disparate disciplines as sociology, urban studies, anthropology and Italian studies.

Gregory Smith is affiliated with various academic institutions in Italy. His interests include Italian rural communities and marginal peripheral urban environments. He has published on farming communities in central Italy and on public space in Rome.

Gilda Berruti is an architect and Associate Professor of urban planning at the Department of Architecture, Federico II University of Naples, Italy. Her research interests include the social construction of urban policies and plans, the sustainable city as an aspect of the new urban question and the relationship between planning rules and urban informalities.

The Landscapes of Italian Food

Local Development and Healthy Practices

Gregory Smith and
Gilda Berruti

LONDON AND NEW YORK

First published 2023
by Routledge
4 Park Square, Milton Park, Abingdon, Oxon OX14 4RN

and by Routledge
605 Third Avenue, New York, NY 10158

Routledge is an imprint of the Taylor & Francis Group, an informa business

© 2023 Gregory Smith and Gilda Berruti

The right of Gregory Smith and Gilda Berruti to be identified as authors of this work has been asserted in accordance with sections 77 and 78 of the Copyright, Designs and Patents Act 1988.

All rights reserved. No part of this book may be reprinted or reproduced or utilised in any form or by any electronic, mechanical, or other means, now known or hereafter invented, including photocopying and recording, or in any information storage or retrieval system, without permission in writing from the publishers.

Trademark notice: Product or corporate names may be trademarks or registered trademarks, and are used only for identification and explanation without intent to infringe.

British Library Cataloguing-in-Publication Data
A catalogue record for this book is available from the British Library

Library of Congress Cataloging-in-Publication Data
Names: Smith, Gregory, 1951- author. | Berruti, Gilda, author.
Title: The landscapes of Italian food : local development and healthy practices / Gregory Smith and Gilda Berruti.
Description: New York : Routledge, 2023. | Includes bibliographical references and index.
Identifiers: LCCN 2022053899 | ISBN 9781032247236 (Hardback) | ISBN 9781032247274 (Paperback) | ISBN 9781003279914 (eBook)
Subjects: LCSH: Food habits—Italy. | Edible landscaping—Italy. | Community gardens—Italy. | Local foods—Italy. | Slow food movement—Italy. | Food supply—Italy—History. | Agriculture and state—Environmental aspects—Italy.
Classification: LCC GT2853.I8 S625 2023 | DDC 394.1/20945—dc23/eng/20221202
LC record available at https://lccn.loc.gov/2022053899

ISBN: 978-1-032-24723-6 (hbk)
ISBN: 978-1-032-24727-4 (pbk)
ISBN: 978-1-003-27991-4 (ebk)

DOI: 10.4324/9781003279914

Typeset in Times New Roman
by codeMantra

Contents

List of boxes		vii
1	Introduction	1
2	Notes on Italy's edible landscapes: Paesaggi Pini Pinoli	17
3	Urban horticulture	42
4	The landscapes of antimafia farms	66
5	Alternative value chains	88
6	Public administration and innovative value chain management	112
	Index	131

Boxes

Box 1	Selected Value Chain Approaches	2
Box 2	EU Paradigm on Agriculture Based on the Principle of Multifunctionality (MFA)	19
Box 3	Five Research Questions Concerning Urban Horticulture	45
Box 4	Nine Research Questions Concerning Antimafia Farms	67
Box 5	Selected Numbers Concerning the Italian Food System	89

1 Introduction

This book is about Italy and the way its citizens respond to the global food industry. It is especially about the grounding of food in the world of physical experience, reversing a commercializing trend which annihilates place. Consider the average American meal which travels 1,500 miles before reaching the dinner table (Pollan, 2009). The diner evidently has no interest in food sources, nor the environmental, economic, social and cultural implications of the food chain. Food is a commodity like any other. Saint Augustine claimed that you cannot love what you do not know: the consumer of commodity food is no food lover. An antidote to this trend is awareness of food as part of a process which has impact on so many factors, including health, landscape maintenance and local development. Italy is a leader in expressing sensitivity for place in food culture, with strong attention to the production process. This perspective places the consumer at the center of the system, transforming disembodied commodities into living parts of an intimate world. The Slow Food movement has led the way through its concept of responsible pleasure, the modern expression of an ancient Epicurean doctrine promoting the informed use of food through a spatially grounded culinary aesthetics (Petrini, 2013). This doctrine is fundamental in sustaining the struggle against the devastation of industrial agriculture and modern commodity food. Eating, after all, is an agricultural act (Berry, 2010).

What links the diner to the landscape, passing through many stages, is the value chain. Value chain (VC) analysis was first proposed in the 1980s as a model for guiding business decisions in an increasingly interconnected global economy. (See Box 1 for a selective summary of different value chain approaches.)

Introduction

> **Box 1 Selected value chain approaches**
>
> **Filiera** – This approach dates to the 1950s, focusing on optimizing physical product flows generally related to the large-scale processing of commodities.
> **Supply Chain Concept** – This adds elements like finance, information, knowledge, strategic interfirm collaboration but still linked mostly to product flows.
> **Michael Porter Value Chain** – Introduced in the 1980s the term value chain, discussed in terms of the production to consumption value chain in the setting of competitive global markets. Concerned essentially with firm-level strategies.
> **Sustainable Food Value Chain** – This is used in economic development especially directed at countries with less advanced economies. Its central dimension is vertical integration or governance, as well has the horizontal integration of networks. Sustainability is an explicit multidimensional performance measure.
> **Food and Landscape System** – This integrates all food value chains within a given territory in a single concept combined with a notion of the landscape as a system integrating all the economic, social and natural factors involved.
>
> Adapted from Neven, David (2014) *Developing sustainable food value chains. Guiding principles.*
> Rome: Food and Agriculture Organization.

As originally used by Michael Porter, who introduced the term, the VC model proposed a distinction between multidomestic activities linked to a single national setting, and global enterprises where competitive factors are linked among nations. Today it is difficult to imagine any context in which even the smallest economic actor can ignore global influences and global outcomes. Food is no exception. The European Union since the 1990s has attempted to protect its territories from the worst consequences of the globally oriented commercial food industry by recognizing that farming does more than yield edible produce. This is part of the multifunctionality concept which has driven EU agricultural policy over the past thirty years. Attention to the multifunctionality of farming is also prominent among environmentalists and quality food advocates. In analyzing the multifunctionality of farming activities, we can identify four groups of functions.

Green functions include landscape and wildlife management, animal welfare, biodiversity, nutrient recycling and the proliferation of carbon sinks. Blue functions concern water management and water quality, flood control, water harvesting and wind energy. Yellow functions refer to rural cohesion and vitality, rural development, preservation of cultural and historical heritages, support for regional identity and the provision of rural services offered to a chiefly non-rural public like agro-tourism and agro-entertainment. White functions concern food security and safety (Van Huylenbroeck, 2007:7).

We believe that multifunctionality can be understood in operational terms through a value chain perspective optimizing positive outcomes while diminishing negative ones. It is a farm to fork concept, a way of thinking about food that goes beyond the product to embrace the full gamut of production factors which start and end well beyond the farm itself. This awareness allows the consumer to back away from viewing food as an isolated product and think about the process in which it is embedded. This perceptual pivot – a Copernican revolution as compared to commercial food orientations – is fundamental if we are to promote positive environmental and health outcomes while rewarding sustainable economic practices. The value chain is not only the way engaged consumers think about food, but also the way socially and environmentally sensitive food producers conceptualize their activities.

A grounded aesthetics putting the physical person in touch with their natural context can bridge the scalar gap between landscape and person. The natural landscape is the source of western aesthetics, today often obscured by the ability of advanced market economies to blind us to our physical constitution (Dear and Flusty, 2000). The consumer has been splintered and separated from their natural condition in the triumph of purely plastic meaning stripped of its natural constraints. The antidote is to restore the synergetic relationship between human actors and the physical world. This involves an aesthetics which is in part a return to the imitation of nature yet goes beyond this conception since nature itself is so heavily shaped by human activity. The aesthetics of nature and human ingenuity are inextricably intertwined (Rancière, 2020).

In healthy conditions a synergy exists between humans and the environment with important implications for personal wellbeing. We are in nature and nature is in us, a relationship governed by myriad factors, not calculating reason alone. Intuition plays an important role in human experience, as does the ordinary practice of everyday life. The world comes into being through a living process known as worlding, enacted through inhabitation rather than being formed in the mind

(Stewart, 2010). This is the intimate world of daily existence, where simple actions create the physical context in which our lives unfold. If we are guided by the world of media representation alone, we impoverish the world and by corollary our own existence: a barren setting for the unfolding of a wretched life. If we direct attention to our physical condition, we live the continuous marvel of nature's mysterious bounty. The natural setting is the landscape, understood as the natural habitat and its built extensions. It is physically and socially constituted, as is the human body itself (Butler, 2010). The landscape and the person have linked trajectories, united by a shared constitution: two dimensions of the intimate. The landscape is a living phenomenon constantly changing and requiring constant attention: etymologically, it is not only the shape of the land but also the process of shaping it.

Humanity's current wayward course is driven by distance from the world of direct experience. Fortunately, awareness of our plight is growing, sharpened by the consequence of the COVID-19 pandemic, fueling concern for sustainability as expressed in personal health, the environment, social inclusion and sound development. In the modern agri-food industry, this distance is the outcome of a commercializing trend rooted in the early postwar years focusing on commodity outcomes and no more. Included among the many negative externalities generated by this approach are the loss of biodiversity, depletion of soil fertility, erosion, water pollution, the loss of small farms and debilitating health implications.[1] There is also the fragmentation of the landscape and the separation between the human subject and the territory. Sound urban and regional planning focuses attention on context. The lack of attention to context is evident in degraded urban and regional environments arising from dwindling public investment and poor planning. The new food equation leverages the multiple benefits of sustainable agricultural practices and healthy diets, while at the same time rehabilitating the physical landscape (Morgan, 2015). This is especially important in urban and peri-urban landscapes which have been splintered by piecemeal planning policies focused on real estate development rather than on the landscape understood as a broad context. Fragmentation is particularly evident in peri-urban landscapes, that ring of land stretching some ten to thirty kilometers from the built city, areas which are now coming under increasing scrutiny (Gisotti, 2012). This is the third landscape, fragmented extensions of the territory neither urban nor rural, defined as the sum of space left over to landscape evolution, to nature alone, a kind of genetic reservoir of the planet, the space of the future (Clement, 2014).

Many contemporary efforts aim to stitch back together these green fragments into a comprehensive system leveraging the creative potential inherent in the territory. We have an excellent example of this in the Agricultural Park of South Milan, a vast extension falling within the peri-urban territory of Italy's most industrialized city. The park was established in 1990, bringing together the built environment, industrial production sites, 1,400 farms, pastures, meadows and woodlands. The park ensures land conservation while creating an ecological balance with the metropolitan area. Agriculture is fundamental in driving this process, supported by the short value chain. One of the most visible park initiatives is coordinated by an association called the Third Landscape, which defines itself as a community of transdisciplinary practices bringing together architects, changemakers, exhibition curators and artists.[2] These activities remind us of the landscape's bursting aesthetic potential.

Vast tracts of fragmented landscapes are found all over the world, appropriately dubbed lost space (Trancik, 1986). Lost spaces are created by policy orientations which focus on single investments rather than looking at the territory understood in comprehensive terms. This is partly a consequence of lessening attention to the public sphere, partly the consequence of dwindling public resources available to support these spaces. Given these political constraints, policies aiming to revitalize lost space must necessarily leverage healthy entrepreneurial practices, ensuring the viability of private enterprises while stimulating non-commodity outcomes that support public interest in landscape revitalization. A fundamental tool in achieving these policy objectives is value chain analysis.

In its original conception, value chain analysis was presented as an alternative to comparative advantage theory (Porter, 1985). Yet though apparently opposed, the two approaches can be usefully combined. The comparative advantage model explores how geography defines the potentialities of a territory, understood in broad terms to include historical accident, the localization of information, the presence of specialized suppliers and specialized workers in what is referred to as thick labor markets (Krugman, 2002). Italian wines are an excellent example of comparative advantage with deep roots in the soil and the climate. Locally grounded agricultural activities all over the country harness the unique potential of single plots cultivated over countless generations. Viticulture's capacity to draw from unique territorial potentiality endows wine with such fine-grained sensitivity as to change character even within the space of a few meters (Scienza et al., 2019).

This remarkable sensitivity constitutes a fascinating area of human endeavor for producers and consumers alike. Of course, all agricultural products benefit from micro-territorial variation, as we see in many cases, such as the black celery cultivated in the Umbrian town of Trevi.[3] Cereals cultivated using domestic wheat varieties is an emerging area of excellence in Italian agriculture, also leveraging micro variations in the territory.[4]

The way to navigate this variety is to follow the value chain. From the standpoint of the consumer such navigation requires knowledge of the production process. Wines are characterized by high added value, making it possible for viticulturalists to invest in unique territorial characteristics. A rich market promotes rich variety, a dynamic which stimulates consumer interest in process. To allow consumers to find their way through this complexity, quality vintners furnish a technical chart illustrating the viticultural and winemaking aspects of their products. But the quality of even low-cost foods, for instance lentils, is directly linked to the production process, making knowledge critical in shaping the choices of quality food consumers.[5] Here we are describing attention to factors shaping product outcomes, but increasing numbers of consumers are also seeking to understand the non-commodity outcomes which are tracked in a multifunctional approach to agri-food systems. These outcomes are of increasing importance in policy orientation, and of notable importance to quality farmers who rightly see themselves not only as commodity producers but also as landscape stewards.

Part of the Copernican revolution in the new food paradigm is the move away from the audit culture that stresses the efficiency of commodity outcomes while ignoring the multiple externalities generated by any economic process. Audit culture is built on a notion of practical reason conceived as a context free quality, an expression of absolute rationality (Strathern, 2000). The fallacy of this conception has long been recognized (Polanyi, 1957; Beckert, 2009). Practical reason thus conceived obliterates cultural difference and annihilates the deep and ever-changing mystery of the territory whose potential evolves through the encounter between nature and human ingenuity, where humans are no longer seen as the subject and nature the object, but rather as partners in an ongoing dialogue (Latour, 2007). Human awareness of territorial factors builds sensitivity to long-term sustainability as well as short-term gain. These factors cannot be subordinated to a strict principle of economic accountability alone and shade off into such unquantifiable realms as appreciation for tradition, or the pure pleasure of living in harmony with nature. Here we must bear in mind Gustav

Mahler's often quoted stricture that tradition is the preservation of the fire, not the adoration of the ashes. The fire is the embedded relation between the human subject and the surrounding territory beyond the financial accounting of the modern agri-food industry. Fortunately, the implied economic rationality of modern audit culture increasingly fails to predict consumer choices. Take the example of olive oil prices in Italy. Italian producers raised a clamor when a European Union agency issued a statement in 2020 asserting that the fair price for extra virgin olive oil (EVO) produced in the EU was €2.28 per liter (Pelagalli, 2020). Anyone who lives in a territory where olive oil is produced knows that this is far below a fair price, based on knowledge generated by direct interaction between producers and consumers. In Italy's central and southern olive growing regions, EVO purchased from a producer – or directly from an olive oil mill – goes for €10 per liter. This is held to be the fair price in the Sabine hills north of Rome, for instance. At this price the local EVO production sells out within a few weeks after the harvest. Buyers know how much it costs to bring a liter of EVO to the market, including labor, plant maintenance, harvest, transportation, pressing and so forth. This knowledge is locally shared, along with an understanding of the nuances of local varietals and cultivation techniques. The consumer knows that these costs make a low price impossible. The consumer also knows that supporting alternatives to the local product will have myriad negative non-commodity outcomes for the territory. This is an example of worlding where affect as well as reason drive actions that support the landscape as a multidimensional component of the human condition.

Grasseni (2009) has demonstrated admirably the tension between local practices and the logic of the audit culture in the production of farmstead cheeses in the Italian Alps. The production of these cheeses is embedded in local systems of knowledge and practices where comparative advantage must be reconciled with concern for the local conception of product quality. The range of factors involved in the local conception are difficult to convey to a remote market which is likely to reduce quality to a generic concept of typical local food. These global structures of common difference (Wilk, 1995) make it difficult for quality producers effectively to communicate their practices beyond the realm of the local market. The producers are aware of this obstacle, for they too are immersed in the global market, and use the value chain to increase the efficiency of commodity outcomes without compromising their traditional concerns.

The comparative advantage of the territory starts with the environment itself, celebrated both in local and national discourse all up and down the peninsula. One of the most famous formulations of Italy's comparative advantage is found in a speech delivered to the Italian parliament in the 1880s, when a deputy from Naples expounded on a comparison between Italy's sun and the fossil fuels which drive the economies of northern Europe (Perry, 2019:88). The question raised was why Italy should betray its natural predisposition, encouraging industries based on imported fossil fuel rather than cultivate the nation's bountiful soil. The sun is the country's natural advantage.

This simplistic formulation glosses over important factors, including localized knowledge systems. Knowledge and practices concerning century-old transformation processes have a weight beyond the physical potentiality of the territory. In many sectors of the Italian economy, expert capacity to process imported materials constitutes a fundamental source of comparative advantage. Not by chance, manufactured products which rely almost entirely on imported raw materials are among Italy's leading exports, like machinery or automobiles. The role of processing expertise is also found in the agri-food sector where imports exceed the value of exports, imports that often constitute the raw material for exported foods. This paradox is made possible by the value chain, where expert processing knowledge can be applied to expertly selected imports to yield foods of rare quality. Inevitably, market segmentation is such that we also find commercial products of low quality, many of which are produced in Italy. Dietary changes are also important here, especially with the increased consumption of meat, which formerly constituted a minor part of the national diet. Today domestically bred and processed meat products dominate the high end of the market, and imports cater to the low.

The highest value Italian exports are prepared agricultural products (PAPs) including wines, cheeses, cured meats, canned tomatoes, olive oil and pasta. PAPs account for 70% of EU agricultural commodities.[6] In most cases, long-established processing expertise is combined with the physical advantages of the territory. Yet some major food exports, like chocolate and coffee, rely entirely on imported raw materials. Processing expertise leverages specialized information, specialized suppliers and thick labor markets, and operates even when isolated from the land's physical potential.

In some cases, food products for which Italy is famous can be made using domestic raw materials, but instead use chiefly imports. Pasta is the best example of this, where the physical characteristics of large-scale producers like the Ukraine, Russia, Canada and the United States allow for economies of scale which are inconceivable in Italy. The lower cost of these raw materials encourages Italian pasta producers to use imports rather than domestic cereals. Historically, Italy's most famous pasta producing area, Gragnano, relied on domestic hard wheat grown along the Adriatic coast to make its pasta. Then, in the early nineteenth century, wheat from the Ukraine was discovered and shipped to Italy in industrial quantities (Falcioni, 2007:24). The interruption of these wheat exports caused by the Russian invasion of the Ukraine in 2022 could take us back to the historical starting point. Higher wheat prices may encourage a return to the use of domestic cereals in pasta production, perhaps even the use of domestic wheat varieties like Saragolla, which was once the staple of Gragnano pasta production. The cost of hard wheat almost doubled between April 2021 and May 2022, and it is inevitable that Italians will have to become accustomed to paying a higher price for their pasta.[7] Yet this richer market could represent an opportunity for Italian wheat producers. The issue is, however, complex since so many variables are involved all up and down the value chain. Before the war, wheat from the Ukraine and Russia accounted for only 5% of Italian wheat imports, so the loss of these imports is not the only factor causing the price increase. The spike in all spheres of agricultural production is linked to soaring energy and fertilizer costs caused by the EU embargo on Russian petroleum products.[8] In the short run, this increase represents a major challenge for Italian agriculture. Perhaps in the future it will represent a boon for organic production. The current crisis in the Ukraine has in any case brought the question of agri-food system to the attention of the entire nation, prompting discussion on all the factors which contribute to bringing a distinctive, healthy and culturally appropriate meal to the table.

Comparative advantage is never absolute, and advantage in one area may be offset by disadvantages in others. The physical advantages of the territory may be offset by high taxation, onerous bureaucratic controls or stringent standards of environmental control. Organized criminality in Italy has long represented a disadvantage in some parts of the territory. Italian products also have a higher cost owing to small-scale quality production. Labor costs are high, creating challenges for such labor-intensive agricultural products as saffron (Leoni et al., 2020). These disadvantages are offset by product quality, and processing expertise transmitted informally from one generation to the next as well as through agricultural training and research. In farming, there is always tension between tradition and innovation, and already in ancient Rome the agricultural essayist Columella aimed to improve farming by promoting such innovative agricultural practices as using organic fertilizers to replenish the soil. Some of his recommendations are still followed today. For instance, the cheese produced in central Italy called *Caciofiore di Columella*, described in his *De Re Rustica*, is a distinguished ancestor of *Pecorino romano*. It has the peculiarity of using a thistle or artichoke flower to coagulate raw milk, making it attractive today for cheese-loving vegetarians who reject the use of animal rennet in cheesemaking. Then as today this cheese is made from the milk of sheep celebrated from ancient times (Armstrong, 2016).

The rediscovery of Columella's works in the Renaissance inspired a revival of agricultural writing which carries on in modern times. The most famous Italian association promoting agricultural excellence through scholarship and modern practice is the Academy of the Georgofili – lovers of the earth – established in Florence in 1753. The founders of the academy were scholars and practitioners who aspired to promote new standards of agricultural production. A Georgofilo of enduring importance is Cosimo Ridolfi who introduced revolutionary terracing techniques into Tuscany in the nineteenth century. As a scholar, Ridolfi was familiar with the ancient agricultural writings, and used their example to inspire contemporary innovation. Columella had assured his readers that the combination of diligence and science would bring economic returns, and among his important contributions to agricultural innovation was that of growing grape varieties separately from one another as is still done today. Ridolfi continued this innovative pathway with the terracing system which not only encountered much criticism in his day but also helped propel the qualitative growth of Tuscan wines (Lambruschini, 1866). Ridolfi realized that his innovative ideas would only be successful if the skills required to implement them were transmitted to workers in specialized training programs. For this reason, he created Italy's first agrarian institute in 1834 at Meleto Val d'Elsa.[9] Today these agrarian institutes are found all over the country as part of the contemporary public school system, helping to ensure the thick labor market which is so important in defining the Italian agrarian landscapes we witness today.

Historical accident is another factor in comparative advantage. One of the most felicitous moments of Italian history concerns the introduction of the winemaking grape in remote peninsular history. This is *Vitis vinifera sativa*, a grape as old as western civilization, subject to constant historical innovation with new varieties and new cultivation techniques (Scienza and Balsari, 2007). A turning point in European viticulture was the introduction of American rootstock in the nineteenth century, adopted to deal with an epidemic caused by the introduction of the Phylloxera pest from America. Grape variety, and wine character, is determined by the scion grafted onto the rootstock, and thus by introducing naturally pest-resistant American rootstock Europe was able to preserve and advance its unparalleled winemaking tradition. A similarly dramatic transformation is now on the horizon, as global warming risks to disrupt this timeless dimension of the European diet. Under current conditions of climate change vineyards are exposed to levels of drought stress never experienced before, and the situation is predicted to worsen. Experts are studying approaches to create new drought resistant rootstock on which to graft historic vines.[10] Naturally, these experts are working with Italian nurseries which are among the world's leading suppliers of grafted vines, another source of comparative advantage.

The Italian landscape that has been celebrated in poetry and painting for over three millennia is the ultimate source of the value chain. The Italian concept of the *filiera* captures the idea of concatenated variables stretching from nature to the final consumer. Typically, this term is translated as the production chain or the supply chain; yet this rendering is narrow as compared to the concept adopted here. The value chain model cultivates awareness of the fact that any food product is more than a free-standing phenomenon, and certainly more than a free-floating expression of commercial value. Market value should arise from interconnected production factors embodying natural potentialities exalted through human ingenuity. Awareness of process stimulates consumer interest in the social, cultural, economic and environmental factors which impinge on commodity creation. As an analytic concept the value chain allows consumers and producers alike to identify specific moments of value creation, tracking their intertwining relationships. It reminds us that product value has its origin and terminus in the environment itself.

Urban agriculture can be read 'through a lens of metabolic rift' operating at different scales, spreading in the empty spaces of post-industrial landscapes (McClintock, 2010). Urban agricultural has a host of multifunctional implications, ranging 'from its attempts to overcome disruptions in ecological cycles to its ability to reclaim public space, re-embed food production and consumption with socio-cultural significance and reconnect consumers with their food and the environment' (McClintock, 2010:193). These practices can contribute to sustainable city development and are classified according to locational dimension and strategic focus (Van Tuijl et al., 2018).

Our approach contains a bias. We advocate a change from market value driven by superficial perception to a knowledge-based economy with an important affective component. We are convinced that raising citizen awareness is essential to promoting local transitions. Social and behavioral aspects of the transition are often underinvestigated and yet so important in promoting circular processes (Jonker and Montenegro Navarro, 2018). We witness this transformation in Italian food markets with the increasing importance of a grounded understanding of food production factors in consumer choices. This historic trend is strengthened by today's inclusionary and environmental sensitivity. It is not a question of commodifying social issues (Tinic, 1997), but rather part of a growing sense of individual responsibility for humankind's health and linkages between landscapes and people.

The present volume documents continuity and change in Italian food markets. Both authors have worked for decades on topics concerned with the management of the territory, which is where this volume belongs. Gregory Smith is a social anthropologist who has worked and published on farming communities as well as on peripheral urban areas. Gilda Berruti is an urbanist whose research interests are mainly concerned with the social construction of urban policies and plans, the sustainable city as an aspect of the new urban question, and the relationship between planning rules and urban informalities. Together we document contemporary trends and illustrate policy measures aiming to accompany a journey toward a more responsible use of the world's resources. The first chapter identifies the general framework for the book. The second chapter establishes interconnections between the landscape, its natural bounty, and the consequences of individual consumer choices. The aesthetics of the landscape carry over into an appreciation of the natural world, and the personal aesthetics of the individual citizen. These widely different scales are strictly interconnected in a seamless expression of value. Chapter 3 explores Italian community gardens, and the multiple policy frameworks which govern these activities. Urban and peri-urban horticulture are part of a long tradition with many social, environmental and health benefits. The drive for locally produced food is a recurring theme in public discussion. While few advocate national self-sufficiency, a solid presence of locally based civic horticulture drives awareness of food production factors, with beneficial impact on patterns of food consumption and food procurement. Chapter 4 considers social farming aiming to contain the negativities generated by organized crime groups. Activists operating on confiscated mafia properties are especially sensitive to the social and environmental dimensions of the value chain. This chapter provides attention to the impact of the COVID-19 pandemic on food-related activities. Chapter 5 looks at the broader context of food procurement, including food cooperatives, buying groups and innovative value chain organization in modern distribution systems. It also explores innovative uses of wheat cultivation and reflects on the implications of the Russian invasion of the Ukraine on Italian wheat production. The final chapter looks at the policy frameworks governing the initiatives described in the earlier chapters, including Milan's Urban Food Policy Pact and Turin's Comprehensive Food Policy.

Acknowledgments

In ongoing research carried out over so many years, it is impossible to enumerate all the sources of help we have drawn from. We wish however to express our gratitude to the many activists, administrators and scholars who have supported our investigations by supplying invaluable insights into their work, as well as the opportunity to participate in the processes in which they are involved. We also express deep appreciation for the work of citizens from all walks of life in the pursuit of concrete endeavors ushering in inclusivity and deeper sensitivity to the sustainable needs of a changing planet. We also express gratitude to our respective institutions for multiple forms of support for a project which has taken up much of our time over the past decade or so. Finally, we thank our students for their enthusiastic interest in the manifold ways we can all contribute to the creation of a healthier world.

Notes

1 https://www.ucsusa.org/resources/hidden-costs-industrial-agriculture (Accessed October 15, 2021).
2 https://terzopaesaggio.org/#top (Accessed June 30, 2022).
3 https://www.fondazioneslowfood.com/it/presidi-slow-food/sedano-nero-di-trevi/ (Accessed June 30, 2022).
4 In Viterbo, Claudio Pagliaccia has been experimenting over the past twenty years with local seed varieties stressing differences in yield from one area of the territory to another. http://www.tusciawelcome.it/it/news/montefiascone-azienda-agricola-fornovecchino-di-claudio-pagliaccia_271.htm (Accessed June 30, 2022).
5 There are infinite examples of this, such as the black lentils grown in the Enna Hills of Sicily. https://www.fondazioneslowfood.com/it/presidi-slow-food/lenticchia-nera-delle-colline-ennesi/#:~:text=Il%20Presidio%20Slow%20Food%20%C3%A8,dei%20terreni%20per%20la%20coltivazione (Accessed July 1, 2022).
6 https://ec.europa.eu/growth/sectors/food/processed-agricultural-products/definition_en (Accessed October 16, 2010).

14 *Introduction*

7 https://www.ismea.it/flex/cm/pages/ServeBLOB.php/L/IT/IDPagina/11794#:~:text=In%20particolare%2C%20i%20listini%20del,mesi%20di%20novembre%20e%20gennaio (Accessed July 5, 2022).
8 https://www.agi.it/economia/news/2022-03-19/guerra-ucraina-potrebbe-cambiare-agricoltura-italiana-coldiretti-16056541/ (Accessed July 5, 2022).
9 https://www.treccani.it/enciclopedia/cosimo-ridolfi_%28Dizionario-Biografico%29/#:~:text=RIDOLFI%2C%20Cosimo.,e%20ben%20dotata%20di%20terre (Accessed June 18, 2022).
10 https://www.lavinium.it/viti-piu-resistenti-con-un-minore-quantitativo-dacqua-la-risposta-della-scienza-ai-cambiamenti-climatici/ (Accessed July 5, 2022).

Bibliography

Armstrong, Philip (2016) *Sheep*. London: Reaktion Books.
Beckert, Jens (2009) 'The great transformation of embeddedness: Karl Polanyi and the new economic sociology,' Discussion paper, No. 07/1. Cologne: Max Planck Institute for the Study of Societies.
Berry, Wendell (2010 [1990]) *What are people for? Essays*. Berkeley, CA: Counterpoint Press.
Butler, Judith. 2010. *Frames of war: when is life grievable?* London: Verso.
Clement, Gilles (2014) *Manifest du tiers paysage*. Paris: Sens et Tonka.
Columella, Lucius Junius Moderatus (1996) *De re rustica*. Edited and translated by Francesca Boldrer. Pisa: Edizioni ETS.
Dear, Michael J. and Steven Flusty (2000) 'Postmodern urbanism,' *Annals of the Association of American Geographers*, 90:1:50–72.
Falcioni, Paola (2007) 'Vocazioni ambientali e tradizioni culturali nella produzione della pasta. Gragnano e Fara San Martino,' *Geostorie. Bollettino e notiziario del centro italiano per gli studi storico-geografici*, 15:1:23–39.
Gisotti, Maria Rita (2012) *Paesaggi periurbani. Lettura, descrizione, progetto*. Florence: Firenze University Press.
Grasseni, Cristina (2009) *Developing skills, developing vision. Practices of locality at the foot of the alps*. Oxford: Berghahn Books.
James, Paul and Manfred B. Steger (2014) 'A genealogy of globalization: the career of a concept,' *Globalizations*, 11:4:417–34.
Jonker, Jan and Naomi Montenegro Navarro (2018) 'Circular city governance: an explorative research study into current barriers and governance practices in circular city transitions across Europe,' European Union, European Circular Economy Stakeholder Platform. https://circulareconomy.europa.eu/platform/sites/default/files/circular-city-governance-an-explorative-research-study-into-current-barriers-and-governance-practices-in-circular-city-transitions-across-europe-2018.pdf (Accessed August 22, 2022).
Krugman, Paul (2002) 'Ricardo's difficult idea: why intellectuals don't understand comparative advantage,' in Paul Krugman (ed.) *The economics and politics of international trade*. London: Routledge, pp. 40–54.

McClintock, Nathan (2010) 'Why farm the city? Theorizing urban agriculture through a lens of metabolic rift,' *Cambridge journal of regions, economy and society*, 3:191–207. https://doi.org/10.1093/cjres/rsq005.
Lambruschini, Raffaello (1866) *Elogio del presidente march. Cosimo Ridolfi*. Florence: Cellini.
Latour, Bruno (2007) *Reassembling the social: an introduction to actor-network-theory*. Oxford: Oxford University Press.
Leoni, Valeria, Luca Giupponi, Davide Pedrali, Giulia Ceciliani, Stefano Sala and Annamaria Giorgi (2020) 'Indagine sulla produzione e la qualità dello zafferano in Italia,' in Congresso della Società Botanica Italiana. Florence: Società Botanica Italiana.
Morgan, Kevin (2015) 'Nourishing the city: the rise of the urban food question in the Global North,' *Urban studies*, 52:8:1379–94.
Pelagalli, Mimmo (2020) 'Olio Evo, per AGEA il prezzo giusto è 2,28 al litro,' *Agrarie notizie*, August 25. https://agronotizie.imagelinenetwork.com/agricoltura-economia-politica/2020/08/25/olio-evo-per-agea-il-prezzo-giusto-e-228-euro-al-litro/67748 (Accessed December 8, 2020).
Perry, Arthur Latham (2019) *Principles of political economy*. Glasgow: Good Press.
Petrini, Carlo (2013) *Slow food nation: why our food should be good, clean, and fair*. New York: Rizzoli Publications.
Polanyi, Karl (1957 [1944]) *The great transformation*. Boston, MA: Beacon Press.
Pollan, Michael (2009) *The omnivore's dilemma: the search for a perfect meal in a fast-food world*. London: Bloomsbury Publishing.
Porter, Michael E. (1985) *Competitive advantage. Creating and sustaining superior performance*. New York: Free Press.
Rancière, Jacques (2020) *Le temps du paysage: Aux origines de la révolution esthétique*. Paris: La Fabrique.
Scienza, Attilio and Paolo Balsari (2007) *La vite e il vino*. Milan: Bayer Crop Science.
Scienza, Attilio, Alessandra Giorgianni, Diego Tomasi, Federica Gaiotti, Federico Graziani, Luigi Mariani and Paolo Carnevali (2019) *Atlante geologico dei vini d'Italia: vitigno, suolo e fattori climatici*. Florence: Giunti Editore.
Stewart, Kathleen (2010) 'Afterword: worlding refrains,' in Melissa Gregg and Gregory J. Seigworth (eds.) *The affect theory reader*. Durham, NC: Duke University Press, pp. 339–53.
Strathern, Marilyn (ed.) (2000) *Audit cultures*. London: Routledge.
Tinic, Serra A. (1997) 'United colors and untied meanings: Benetton and the commodification of social issues,' *Journal of communication*, 47:3:3–25.
Trancik, Roger (1986) *Finding lost space: theories of urban design*. Hoboken, NJ: John Wiley and Sons.
Van Tuijl, Erwin, Gert-Jan Hospers and Leo Van Den Berg (2018) 'Opportunities and challenges of urban agriculture for sustainable city

development,' *European spatial research and policy*, 25:2:5–22. http://doi.org/10.18778/1231-1952.25.2.01.

Van Huylenbroeck, Guido, Valerie Vandermeulen, Evy Mettepenningen and Ann Verspecht (2007) 'Multifunctionality of agriculture: a review of definitions, evidence and instruments,' *Living reviews in landscape research*, 1:3:1–43.

Wilk, Richard (1995) 'Learning to be local in Belize: global systems of common difference,' in Daniel Miller (ed.) *Worlds apart: modernity through the prism of the local*. London: Routledge, pp. 110–33.

2 Notes on Italy's edible landscapes

Paesaggi Pini Pinoli

This chapter explores the powerful interconnections between ordinary citizen actions and the physical world around us. The idea draws from Bruno Latour's (2017) suggestion that the world is not made up of an opposition between social beings and nature; nature has a social character, and humankind has a natural side. His idea of actants is particularly relevant, where he shows that humankind is not always the active force in engagements with nature: nature can also take the initiative. Beauregard (2012) notes that human activity always occurs in conjunction with non-human things: the world must be understood in its materiality. We see this in climate change, where nature has reacted to human interventions by dictating conditions to which the human community must respond. This is seen in unpredictable weather patterns, with heavy precipitation followed by long drought, and strong alternation between high and low temperatures. These patterns create real challenges for plants and farmers alike. Today the Italian press uses the term 'water bomb' to describe heavy cloudbursts with a rain rate of greater than 30 mm per hour. In intense such phenomena 100 mm of rain may fall in three hours, an enormous amount considering that in many parts of central Italy the annual rainfall is only 700 mm. Such atmospheric calamities reveal the interdependency of humankind and nature, with nature taking the role of the initiator. Nature demonstrates its agency in a complex social scheme where humans play at times the role of subject, at times object. As the years pass, the situation becomes ever more dramatic, and 2022 recorded the most severe drought in seventy years creating conditions where rivers literally dried up, and Italy's largest river, the Po, flowed at only 20% capacity.[1] In these conditions, everyone realizes that the relationship between the human community and the natural environment needs to be rehabilitated.

The choice of a chapter title in Italian provides the pretext to show aesthetic interconnections between place and language. The poetic

DOI: 10.4324/9781003279914-2

technique of alliteration, repeating identical initial consonants, is a poetical device based on the physical character of words. All realms of the physical world contain constraints which are also opportunities for cultural elaboration. We are masters of nature but bow to natural constraints. Parataxis is another poetical device with a long tradition, in which words are juxtaposed with no grammatical indication of the relationship between them. No element is more important than any other, no hierarchy is implied. This expresses a democratic relationship between the human subject and the surrounding world.

Our ability to perceive the world has been strongly impacted by a system of commercial representation which atrophies our capacity for perception. This is what Guy Debord talks about in *The Society of the Spectacle* when he claims that the market system has turned place into a site of commodity consumption. Today we are confronted with a simplified two-dimensional vision of the world which denies us the pleasure of deep visual exploration. We have a natural proclivity to view any open space, upon our first gaze, in a manner which reveals only a two-dimensional form allowing the viewer to establish a sense of order (Kaplan et al., 2012:45). Within milliseconds, however, the gaze shifts automatically into a three-dimensional mode, allowing the viewer to experience depth and mystery. It is precisely this innate visual curiosity that has been mortified by a commercial sign system. Our capacity to see impacts our ability to form a mental image of the world around us: an impoverished language of visual experience blunts the world as it presents itself to our gaze. Too many citizens can instantly recognize a commercial sign but are unable to distinguish between a pine and a plane tree; indeed, they are not much interested in the distinction. One way to reverse this deficiency is to offer training on environmental education. Among the most prestigious such courses are those entitled to Emilio Sereni, like the schools sponsored by the Cervi Institute in Emilia Romagna, or the Coraggio Cooperative in Rome.[2]

Modern convention encourages us to believe that nature exists somehow 'out there,' removed from the realm of personal interest and responsibility. But it is increasingly recognized that the world is the way it is because we are in it. And we are the way we are because nature is in us. Yet the relationship is far from symmetric, since nature can exist without us, but we cannot exist without nature (Latour, 2017). This is seen in the ramification food consumption has for our wellbeing, for plant and animal life, for the physical environment which envelops us. We draw from nature in our cultural constructions, and our cultural constructions then play back into our appreciation of the world around us. The gaze is culturally constructed action based on an aesthetics which informs our every action.

Notes on Italy's edible landscapes: Paesaggi Pini Pinoli 19

The ideas explored here borrow significantly from the concept of agricultural multifunctionality. This concept gained currency in the 1990s and since then has governed EU agricultural planning policy (See Box 2 for selected approaches to multifunctional agriculture).

Box 2 EU paradigm on agriculture based on the principle of multifunctionality (MFA)

Nature of agriculture

- Commodity outcomes from farming are not sufficient to maintain the economic viability of firms contributing significantly to multifunctional outcomes.
- Non-commodity outcomes have high economic, social and environmental value. These outcomes constitute underrewarded public goods.

Policy objectives

- Preserve the rural landscape as a multidimensional public good.
- Maintain the viability of small-scale farming operations in virtue of their greater multifunctional capacity as compared to large-scale operations.

Policy instruments

- Subsidies to farms in virtue of their contribution to landscape management.
- Protection of small-scale farming enterprises against commercially oriented monofunctional agriculture.
- New institutional arrangements allowing firms to optimize their multifunctional outcomes.
- Rural development plans supporting farming and forestry activities which promote and protect the landscape.

Adapted from Van Huylenbroeck, Guido, Valerie Vandermeulen, Evy Mettepenningen and Ann Verspecht (2007) 'Multifunctionality of agriculture: a review of definitions, evidence and instruments,' *Living reviews in landscape research*, 1:3:1–43.

20   *Notes on Italy's edible landscapes: Paesaggi Pini Pinoli*

The Italian landscape and visual aesthetics

Live drawing is an inadequate translation of the Italian *disegno dal vero*, literally drawing from the truth inherent in the physical world around us, both in the sense of borrowing and representing. This activity would have been standard practice for any art student up until a few generations ago. By representing nature in graphic form, the artist cultivates an appreciation for the infinite ways the world reveals itself to us, celebrating nature while establishing a standard for human aesthetic achievement. The discipline of landscape drawing reaches far back in western history, as we see in the stunning paintings from Livia's dining room excavated in Rome in the 1950s. This set of frescoes exalts the consumption of food products within the pictorial representation of nature recreated within a Roman suburban villa; a multilayered narrative encased in a delightful mise-en-scène.

Emilio Sereni's magisterial *History of the Italian Agricultural Landscape* is the foundation text for anyone interested in Italian landscape semiotics. This discipline allows us to decipher the meaning expressed by an open viewscape, in Italy almost invariably connected to the country's noted built achievements. Ambrogio Lorenzetti illustrated in the early fourteenth century the strict interconnection between the city and the countryside in his celebrated fresco cycle on the consequences of good and bad governments in Siena. Cities and the countryside have always risen or fallen in a single cycle.

Ancient Roman principles of landscape organization are still visible in many parts of the Italian territory. The urban principles drawn from the Greeks were extended by the Romans to peri-urban areas, endowing the landscape with regular geometric forms. Regular fields given over to cereal cultivation, typical of Roman plantations, prevailed in the central and northern parts of the peninsula, while in the southern Italian areas irregular plots devoted to horticulture were common. The latter were typical of the Mediterranean garden, with trees and vines surrounding scattered buildings. These dwellings could be occupied by farm workers and farm managers, then called a *villa rustica*, or they could be a place of pleasure (*locus amoenus*) built for the farm owner, then called a *villa suburbana*.

Plantations became firmly established in the late Republic, mostly devoted to cereals grown in a two-field system where cultivation alternated with fallow. For Romans, the triumph of law and order was expressed through the presence of a regular landscape based on the *centuria*, a geometric shape containing about fifty hectares. This form is still visible in many parts of Italy, such as the area north of

Bologna which has recently been the focus of political debate owing to a planned beltway that would destroy this ancient feature of landscape heritage.[3] The Roman system, with its emphasis on order and regularity, separated agriculture and animal husbandry in a closed field system where pasture for animal husbandry had to be made available outside the plantation itself. Thus, in an early period public grazing (*pascua publica*) and woodlands (*silva*) were inserted by landscape surveyors into the Roman countryside, complementing the regularity of cultivated fields. Still today pastures represent an important part of the landscape, protected under EU policy.[4] Today, semi-natural permanent pastures represent a source of biodiversity drawing from variations in microclimate, altitude and soil characteristics, while also constituting an important carbon sink.[5] Semi-natural woodlands are likewise defended under EU policy.[6]

As Rome grew in sophistication so developed a concern for landscape aesthetics which competed with the practical value of farmland. Indeed, Varro writing in 37 B.C. urged that farms should be beautiful but also rational and useful. The landscape factor of the *bel paesaggio* is still important today, with significant economic benefits which go beyond agriculture itself. Tuscany's famous variety of landscapes is a powerful magnet for tourist flows. In some parts of Europe, effort has been made to create financial transfers from non-farm to farm sectors in areas whose landscape value has benefited markedly from the contribution of farms (Van Huylenbroeck et al., 2007:26).

Italian regional difference

Italy is a nation one and indivisible, yet simultaneously promotes the autonomy of its different territories. Having only been united as a nation in 1861, it is inevitable that the variety of landscapes is matched by enormous cultural variety. A measure of cultural variety is the fact that no Italian language existed until its artificial creation in 1868 as a mandate to promote an element of national uniformity (De Mauro, 2011). Yet localism has powerful roots in Italy and dialects continued to prevail over language throughout most of the twentieth century. Extreme micro variation in speech was match by difference in all cultural spheres, not least the culinary realm. Linkages between food and speech were once expressed by the invitation to 'talk like you eat' when in the local realm a townsperson put on cosmopolitan airs by speaking the national idiom rather than the local dialect. Effort to organize culinary variety into a national cuisine dates to the 1891 publication by Pellegrino Artusi of *Science in the Kitchen and the Art of Eating Well.* Rather than using his own Romagnolo dialect in this publication, Artusi used Tuscan as the most plausible candidate for a national language that could harmonize the nation's extraordinary variety. Italian culinary variation is extremely fine grained, with changes from one town to the next belying any idea that Italy's cuisine is only regional. It is in fact local, but with territorial affinities that make it possible with some qualification to speak, as does Artusi, of regional culinary traditions. This fine-grained variety involves food preparation techniques and is also rooted in variations in the natural potentiality of Italy's many landscapes.[7]

In climatic terms, the national territory is divided between Alpine, continental and Mediterranean areas. Geographical distinctions can be further broken down into an enormous range of meso regions and micro territories concerning a host of variables. Broadly speaking, Alpine areas involve specialized agricultural activities suited to a short growing season. These areas are protected under the Alpine Convention of 1991 which promotes the sustainable development of the Alps. As defined by the Convention, the Alpine range stretches 1,200 kilometers through eight states with a maximum width of 300 kilometers between Bavaria and Northern Italy. Austria and Italy together represent more than 55% of the Convention area and with France cover three-quarters of the total surface of the Alpine Convention territory. Humans have maintained a balance with these extreme environmental conditions for thousands of years, yet today climate change is playing havoc with the Alpine range, especially seen in the melting of glaciers. The Alps today receive less snowfall than ever before, and this has dramatic impact not only for the Alps themselves but also for the fertile plains of the Po Valley, the richest in Italy and among the richest in Europe. The Po Valley with its deep topsoil, excellent irrigation and continental climate, is not only a flourishing agricultural district, it is also Italy's industrial heartland. These factors ostensibly represent territorial advantages, yet the Po Valley is also one of the most polluted parts of Europe. South of this continental climate area, following the line which stretches across the Apennines from Rimini to La Spezia, we find territories falling broadly within the Mediterranean-type climate area. One of the many challenges faced by these areas are forest fires, prompted by climate change and aggravated in some parts of Italy by organized criminality. Forest fires are an especially serious threat in Sicily, where between 2009 and 2016 fires decimated one-third of the island's total surface area.[8] Three quarters of these fires were caused by human action. The two regions in terms of the numbers of fires which follow Sicily are Campania and Calabria, all three of which are noted for endemic organized criminality. Some shepherds use forest fires as a way of clearing fields so they can serve the grazing needs of their flocks. Fires are also used to clear forest land that can then be used for farming. A third cause is called the fire industry, where unscrupulous citizens start forest fires so they can find employment in reforestation. This destruction of a common good has not been taken seriously by these regional governments.

Climate is not the only variable in shaping the landscape, geology also plays a role. Specific soils are suitable for different types of ground cover creating fine-grained variation in agricultural activities. To take an example from Italy's continental area, we note that the success of Italy's wet rice production in the Vercelli Plain draws from major rivers descending from the Alps. This extensive plain slopes to the southeast where it drains into the Po River. Once entirely woodland, the plain was reclaimed and transformed into farmland starting in the eleventh century. Continual interventions have characterized this area, including the building of Italy's third largest canal, the Cavour Canal, in the nineteenth century. Thanks to all these works Italy is today the most important rice producer in Europe, with exports all over the world. The lucrative character of this market has encouraged some farmers to extend their operations into micro areas that are not suited to this type of cultivation. This is especially the case of areas in proximity of the Po River, where highly percolating gravel soils lose their fertility when given over to intensive agriculture (Regione Piemonte, 2017:156). This extension of wet rice cultivation destroys natural woodlands, leads to greater pollution of the area's shallow aquifer and results ultimately in the desertification of the soil. Desertification is defined as the depletion of once fertile soil to the extent that it loses its capacity to create biomass and is thus unsuited to agriculture using traditional methods (Iannetta, 2006). This is the result of excessive use of heavy mechanical equipment, excessive use of fertilizers and irrigation. These are among the many factors contributing to the pollution of the Po Valley.

The Italian landscape is protected under Art. 9 of the 1948 Constitution, today falling under the authority of Italy's twenty regional governments which were given significant autonomy in the 2001 constitutional reform. All regions are required to devise a landscape plan, and five regions have complied so far: Emilia-Romagna, Lazio, Puglia, Sardinia and Tuscany. Other regions are also making headway in developing a landscape plan, such as Campania Region which approved a provisional plan in 2019.

Regional territorial plans have their roots in the innovative urbanistic law of 1942 which required all towns to adopt master plans, and all regions to adopt territorial plans. This was intended chiefly as a way of coordinating urban growth but has changed over the years, and in 2004 was modified to include concern for landscape planning. This is the Code for the Preservation of Heritage and the Landscape which includes provisions for management and protection, and speaks of the rehabilitation of landscape values through the requalification of degraded areas. These regional laws safeguard landscape values, minimizing the consumption of the territory while promoting urbanistic and building growth compatible with rural landscape values.

Regional planning often encounters strong political opposition owing to the difficulty of reconciling competing territorial interests. We had a case of this in the Veneto Region when the Conegliano-Valdobbiadene Hills were named a UN World Heritage Site in 2019. This recognition was awarded based on a long history of excellence in the spontaneous management of a landscape which is linked to the production of Italy's finest Prosecco wines. Here is the justification:

> Located in north-eastern Italy, the property includes part of the winegrowing landscape of the Prosecco wine production area. The landscape is characterized by 'hogback' hills, *ciglioni* – small plots of vines on narrow grassy terraces – forests, small villages and farmland. For centuries, this rugged terrain has been shaped and adapted by man. Since the 17th century, the use of *ciglioni* has created a particular chequerboard landscape consisting of rows of vines parallel and vertical to the slopes. In the 19th century, the *bellussera* technique of training the vines contributed to the aesthetic characteristics of the landscape.

The concern expressed in 2019 by many environmentalists was that formal acclamation of the landscape's viticultural merits would be an incentive to extend the commercial use of vineyards in a form of monoculture which would destroy the harmony of the landscape by upsetting the balance between woodlands and agricultural pursuits (Gentilini, 2019). Yet the UNESCO mandate rules out the construction of new production areas, and the Veneto Region, as all regions in Italy, has provisions which protect natural green areas, including woodlands, river corridors and variegated sites mandated by the EU.[9]

Small-scale natural woodlands are important throughout the peninsula, as are the abundance of national parks which constitute about 5% of the nation's surface area. Protected green areas also include marine reserves, natural state reserves, protected areas of national interest, regional parks and regional nature reserves. Protected areas

Notes on Italy's edible landscapes: Paesaggi Pini Pinoli   25

mandated by the EU and managed by regions include Sites of Community Importance (SCI) and Special Areas of Conservation (SAC). While there is considerable variety in regional policies, the overall outcome at the national level in promoting green is positive. A measure of success is found in the extension of Italy's forests, which expanded by 20% in the last thirty years. In Europe, Italy is now second only to Spain in terms of forests which cover 38% of the national territory.[10]

Land use and land take

The land take involved in urban sprawl continues to be a matter of concern, occupying as it does seventy square kilometers of Italy's green space each year, equivalent to nineteen hectares a day (ISPRA, 2022). Yet notwithstanding this negative trend Italy's tree cover continues to grow, as does the area given over to agriculture, reversing the European pattern. Between 2009 and 2019, the utilized agriculture area (UAA) in the EU declined by 0.8% but increased in Italy by 4.1%. Today 42% of the national territory is utilized agricultural area, about half being arable, the rest pastures, meadows and commercial forest (Coldiretti, 2019). Commercial forests play a significant role in these trends, since they are increasingly adopted on farmland owing to low labor requirements and easy adaptation to climate change. Indeed, in this same period, woodlands increased extension while the area devoted to cultivated plants declined by 2.9%. Among cultivated plants both hard and soft wheat production increased, while corn declined (ISTAT, 2021). Rice production is also up, and today Italy produces 53% of the EU's rice.[11] The trend is of course not uniform across regions. The increase in wheat production achieved in 2021 was especially thanks to trends in the Italian south. There are also sectoral differences, and permanent crops – in particular vines, fruit and nut trees – in percentage terms occupy more surface area than the EU average, accounting for 20% of all agricultural land. Permanent meadows and pastures have instead a lower incidence than the EU average.

As an environmental trend, all this seems positive unless we look at a deeper historical perspective and see that Italy has lost 50% of its arable since 1960. Many producers claim that this downward trend is the outcome of shortsighted EU planning aiming to leverage cheap imports over domestic products. These policies have penalized Italian agriculture for decades.[12] Many question whether this orientation truly serves the needs of the Italian community, and note that the global market for agricultural commodities is anything but rational, departing regularly from the classical laws of supply and demand. The baffling paradox for producers in the summer of 2022 was the increase in food prices matched by a *decrease* in some agricultural prices. This was especially seen in the case of wheat prices which after a period of increases declined in July of 2022 notwithstanding prolonged reports of how the Russian invasion of the Ukraine would lead to a price hike. That month forecasts were released of a bumper hard wheat harvest in Canada, the world's largest producer of this commodity, causing Italian hard wheat prices to plummet by 15%.[13] Yet Italian public opinion has long been skeptical of Canadian wheat knowing that high yields are made possible due to the heavy use of chemical substances, including carcinogenic glyphosates. About the same time as the bumper crop was announced with its negative impact on wheat prices, a US government study showed that 80% of Americans have traces of glyphosates in their urine.[14] Italian farmers are not allowed to use glyphosates in the same way as Canadian producers, and this should justify higher domestic prices. But the mainstream market continues to allow domestic industrial food processors to use cheap imports which damage Italian public health as well as Italian farmers, not to mention the multifunctional outcomes determined by the role of farmers as landscape stewards.

Woodlands, pastures and meadows have played a substantial role over many centuries in allowing human communities to survive in high Alpine areas (Cole and Wolf, 1974). Still today, in the region of Piedmont, Alpine pastures make up about a third of all agricultural land, of which some 40% are publicly owned.[15] These areas are no longer the source of livelihood for resident communities engaged in subsistence agriculture and are instead promoted and protected in the service of a broader range of outcomes. The regional management plan for meadows and pastures adopted in 2009 (*Piano pastorale foraggero*) aims to ensure landscape preservation, environmental conservation, quality food production, calamity reduction, the augmentation of biodiversity and support for rare livestock breeds. These measures preserve the landscape, as well as selected economic activities, such as the animal husbandry required to produce a range of stunning cheeses. The most remarkable of these dairy products is violet cheese (*Plaisantif*) produced in the Chisone Valley to the northwest of Turin. The milk used to produce this cheese is taken from cows grazed on rich polyphytic pastures which between June and July include violets transmitting their delicate aroma to the cheese. The production techniques are ancient, often involving autochthonous breeds of cows, including Abondance, Pezzata Rossa and Pustertaler.[16]

Polyphytic pastures are found throughout the national territory, each with a specific botanical character determining unique landscape features and dairy products. In central Italy, the Colfiorito Valley is famous for its explosion of spring colors, transformed into peerless *pecorino* cheese which makes exclusive use of local sheep milk. Further south, in Sicily, we find the polyphytic grasslands of the Iblea mountains, a beautiful landscape famous for its *Ragusano* cheese. Over one hundred species of wildflowers populate the Iblea hills, including mint, sage, lavender and wild oregano. To reinforce the identity between the landscape and the product, this cheese is made exclusively from the milk of a Modicana cow, a breed found only in the area of Sicily from which it takes its name. Today the breed is threatened by extinction owing to competition from more prolific breeds, like the Holstein cow which yields twice as much milk as the Modicana. Today there are only 2,000 Modicana cows as compared to over 25,000 in the 1960s.[17] They survive thanks to regulations that require milk from a specific breed for protected designation of origin cheeses like *Ragusano*.

The linkage between food and landscape is detected in almost any Italian agricultural product, including staples. Cornfields once dominated the northeastern region of the country, yielding corn used to make polenta. Corn began to be cultivated on a large scale in this part of Italy in the seventeenth century to feed the region's burgeoning population, with a greater kilocalorie yield per surface area that any other Italian staple except irrigated rice (Coppola, 1979). Yet today corn imports far outstrip domestic production owing especially to the importance of this commodity in the livestock industry.[18] These imports continue to grow, and today account for half the nation's corn market, as compared to 10% in the year 2000.[19] Italy prohibits the cultivation of genetically modified plants on its territory, yet almost all Italy's corn imports, worth one billion euros, use genetically modified seeds not allowed in this country. These imports are employed as cattle feed, or in some industrial applications like the creation of dextrose, corn starch or corn oil. GMO corn is never used to make Italy's most important corn dish, polenta, partly owing to the public perception of GMOs. This choice is also due to a persisting desire to find linkages between food and the territory, and some polenta producers still turn to specialized Italian corn varieties for their product, like white corn (*mais bioncoperla*), or the *Spin* corn of Valsugana. Both varieties are promoted by Slow Food and help maintain the diversity of corn varieties available on the national territory.

The northwest has historically been specialized in wet rice cultivation, an early Arabic import that grew in agricultural significance around the sixteenth century. Agricultural improvements in the nineteenth century, including the extension of the canal system required for wet cultivation, increased productivity, making Italy Europe's most important rice producer. Innovative plant varieties also helped increase capacity and consolidate Italy's hold on the rice market. One of Italy's newest rice varieties is Venus (*riso venere*) developed in the 1990s as a cross between black Asian rice and Italian domestic white rice. Baraggia is Italy's only PDO rice, grown in one of twenty-two towns situated in the heart of the Vercelli Plain. A PDO since 2007, Baraggia comes in seven varieties, the most suited for Italy's premier rice dish, risotto, being the large-grained Arborio selected in the town of Arborio in 1946.[20]

Wheat has the longest history of all the cereals cultivated in Italy, reaching back to spelt production in ancient times consumed as a versatile dish called *puls*. *Triticum spelta* is still produced today by specialized farmers who preserve Italy's rich seed heritage. Industrial farmers instead concentrate on high yield wheat varieties using seeds available to farmers all over the world. Wheat production is a dramatic example of biodiversity loss, declining from the 291 varieties cultivated a century ago to a handful today (Li Rosi, 2012). Seeds historically have been produced and consumed by farmers, leading to abundant experimentation and significant seed variety. Recent interest in environmental diversity has led to the documentation of what are classed as ancient cereals, referring to varieties produced before the rise of industrial farming in the postwar period. Notwithstanding the name, some of these wheat varieties are in fact quite recent. For instance, from the 1920s through the 1950s, the most common hard wheat grown in Italy was Senator Cappelli, developed in 1915 through the selective breeding of north African wheat varieties by Italy's most famous agronomist of the time, Nazareno Strampelli. It was thanks to this high yield hard wheat variety that Italy was able to achieve national self-sufficiency in its wheat production in the fascist period. Cappelli then fell out of favor in the 1960s owing to the creation of modern varieties with even higher yields. This wheat variety is now enjoying a domestic renaissance owing to its strict connection to the Italian territory and its suitability for pasta production. We are of course describing a niche market dominated by small boutique pasta makers. Yet recently one of Italy's most important wheat producers has returned to Senator Cappelli, used to make pasta sold under the trade name Cappelli.[21] This is Bonifica Ferrarese (BF), today a publicly traded company and the most extensive farming operation in Italy. Their success in promoting an Italian wheat variety for a domestic market that too often values price over quality bears witness to important changes among Italian food consumers.

Other wheat varieties falling under the heading of ancient cereals are truly ancient. The regions most noted for the cultivation of these

Notes on Italy's edible landscapes: Paesaggi Pini Pinoli 29

ancient cereals are Sicily and Tuscany. Some fifty varieties of wheat are cultivated in Sicily today, and most are autochthonous, helping preserve Italy's biodiversity, the typical character of its local diets, and its distinctive variety of landscapes (De Cillis, 2004). In Tuscany, the role of wheat cultivation in maintaining the distinctive character of the landscape has won high visibility through the listing of the Val d'Orcia as one of Italy's World Heritage Sites owing precisely to the role played by agriculture. Its wheat tradition reaches back to Neolithic times, and excavations in this area have found traces of the oldest wheat variety cultivated by humans, *Triticum monococcum* (Bellini et al., 2008). This extraordinary find is a forceful confirmation of the profound historical interconnection between diet, agriculture and the landscape.

The stone pine (*Pinus pinea*)

An exploration of pine trees allows us to understand how a distinctive crop can shape a diet and a landscape, knowing that pine trees produce the nuts which play a significant role in some regional dishes. The tree's lofty canopy and majestic bearing has long been celebrated in the Italian arts, including music. The Pines of Rome is a symphonic poem composed in 1924 by the acclaimed composer Ottorino Respighi, using a simple modal structure to evoke the serene quality of Rome's pastoral hinterland. Respighi rightly chose the stone pine as the symbol of the Roman countryside, and was so attached to this landscape feature that he had a villa built in a pine forest on the northern outskirts of the city, aptly styled *Villa Pini*.

This twentieth-century Roman icon has rather recent roots. It appears that the only autochthonous *Pinus pinea* forest in Italy was that once found in Sicily, the tree having been introduced elsewhere much later.[22] The *Pinus pinea* is omitted in the catalogue of trees Pliny the Elder details in his encyclopedic *Natural history*, although pine nuts find their way into Apicius's cookbook of about the same time. Its virtue in landscape enhancement and the stabilization of fragile soils was only discovered in modern times, especially in connection with reclamation works starting in the seventeenth century. These efforts were made necessary owing to the abandon into which many areas of the Italian peninsula had fallen after the collapse of the empire. Reclamation made it possible to extend farming out toward Italy's Tyrrhenian coastal areas, stretching from Liguria, through Tuscany, and on into Lazio.

In the seventeenth century, the wild abandon of central Italy was a template for European landscape painting captured in the vision of Claude Lorrain. He projected on to this countryside 'a perfect image of the state of innocence,' completely undomesticated and for that reason pure (Wine, 1994:29). Rome's countryside was so utterly abandoned as to be described in 1680 as one of the most 'dispeopled' areas of Italy (Beaven, 2014:81). The British admiration for the 'picturesque' image of the Italian countryside ignored the dismal economic conditions of which it was a product. The term was taken up by Italians as *pittoresco*, used with evident irony to express an image of the perceived beauty deriving from Italy's backward state (Gisotti, 2008). This abandon fueled the food insecurity which plagued many parts of Italy and spawned what was described in 1740 as 'a horrid thing called the malaria' (Beaven, 2014:85). This scourge eventually spurred prodigious reclamation efforts adopted as an expression of progress, a way to establish order in the natural landscape (Cavallo, 2011:26).

The reclamation of wild swampland could increase property value by even fifty times, yet the cost was exorbitant, often exceeding the benefits accrued through property appreciation (Cavallo, 2011). To propel the process of rural regeneration public funds were needed. A turning point was in 1882 when the Baccarini law defined land reclamation as an activity of fundamental public utility. Under this law, the imperative of public health forced landowners to undertake reclamation even against their will, and non-compliance led to expropriation. This was a national law with effect all over the country. But the most striking impact was in the area around Rome where enormous and largely inactive estates were fragmented and sold to investors using state-secured mortgages for rural development. New consortia of landowners were created to oversee the drainage works, called *Consorzi di Bonifica* (Cavallo, 2011:33). One such company is still present on the eastern periphery of Rome today, with farm buildings, now publicly owned, known by the name of the historic reclamation company, A.L.B.A. (Anonima Laziale Bonifica Agraria). It is engulfed in an enormous pine forest.

In 1910, a pioneering law was passed promoting what would later become known as integrated land reclamation. It not only required that land be drained but also developed as part of a viable agricultural system. One requirement was that trees be planted, almost invariably the high-canopied pines which dot the landscape in many parts of central Italy, with farmhouses dating to this period connected by networks of country roads lined with magnificent pines. This arboreal presence monumentalized pathways in a harmonious integration of natural and human elements. Today in many parts of central Italy these tree-lined lanes are often interrupted by the massive growth of the cities and modern infrastructure. If one visits Respighi's *Villa Pini* today, the visitor will be confronted by a monumental home now in ruins, and hardly a vestige of the pines which once gave the *Villa* its name. This is the outcome of a long process which transformed wild landscapes into agricultural fields, later abandoned to unsightly urban sprawl.

Nut production in Italy today

The maintenance of a pine tree is simple. If there is pedestrian passage below, periodic pruning of branches is required to prevent them from falling heavily to the ground with attendant risks for possible passersby.[23] This sounds trivial yet must be performed every five to ten years at the cost of some €600 per tree.[24] More important for agriculture, however, is the requirement that trees be cut to the ground every eighty years so a new shoot can grow and bear new fruit (Regione Toscana, 1998). Yet landscape laws forbid cutting these trees down without authorization which is rarely granted since they are an important feature of the visual landscape. In this way, landscape trumps agriculture, contributing significantly to Italy's declining pine nut yields, further diminished by acid rain and pest infestation worsened by global warming.[25]

Italians consume one thousand tons of pine nuts a year. The region of Tuscany alone, the country's leading pine nut producer, harvested 2,000 tons in 1960 (Regione Toscana, 1998). In recent years, the pine nut harvest has diminished by as much as 90%, and with persisting high demand for Italian pine nuts the price is steep. A single pine tree yields about 1.2 kilograms of pine nuts a year, and a hectare of pine forest comprises between 120 and 250 trees.[26] The wholesale price per kilogram is about €50, equivalent to a per hectare value of as much as €12,000. Were it not for a recent pest calamity this would still be a highly lucrative forest product.

The calamity in Italy and the Mediterranean is infestation by an American pest, the western conifer seed bug (*Leptoglossus occidentalis*), for which there is still no treatment. This pest arrived in 1999 and in three years led to a 90% decrease in pine nut yields.[27] Italy was once self-sufficient in terms of its pine nut requirements, yet today one finds many Spanish and Turkish imports. Asian pine nuts are cheaper and can be identified by their elongated shape and brown extremities. These nuts are not as desirable as the Mediterranean fruit owing to their bitter aftertaste. Food retail packaging requires the indication of origin, and Italian pine nut producers will invariably indicate their country as the origin. Mediterranean pine nuts in general are identifiable by their uniform light color and smooth taste; their origin is also indicated.

Dishes using pine nuts

Pine nuts have an intriguing role in binding the landscape to individual dietary choices, since many Italian dishes feature pine nuts, the most famous internationally being pesto. Pesto is the Italian word for a mash which can be produced using various substances. Thanks to pesto, pine nuts are thought of as a condiment for savory dishes, but they are also used in sweets. Pine seeds have a protein content of 31.6%, fat of 44.9% and an energy value of 583 kcal/100 g. Regular consumption of pine nuts is associated with reduced risk of coronary heart disease (Mutke et al., 2005). Their high nutritional value has been appreciated since Paleolithic times.

In ancient Rome, pine nuts were considered a delicacy. Apicius, the famous first-century A.D. gastronome, recorded various recipes using this ingredient. One is soft-boiled eggs in pine nut sauce, a recipe calling for eggs, black pepper, soaked pine nuts, honey, vinegar and a salty condiment called garum (Faas, 2005:316). A modern Roman specialty is dried cod (*baccalà*) with pine nuts, calling for dried cod, peeled tomatoes, onion, garlic, pine nuts, raisins, orange zest, bay leaf, salt and black pepper.[28] Pine nuts and raisins combined produce a mild sweet and sour effect. To move to another region, Sicilian *caponata* also expresses a sweet and sour effect in an eggplant dish which brings together tomato pulp, celery chunks, basil, raisins, pine nuts, capers, vinegar and sugar.[29] These are all savory dishes, yet sweets also abound. The classic is Tuscan Grandmother's Cake (*Torta della nonna*), a simple custard in a pie crust with pine nuts sprinkled on top.[30] Pellegrino Artusi (2003) includes this in his historic 1891 collection of recipes.

The most famous culinary use of pine nuts is in Ligurian pesto. Liguria has a vast coastline, much of which given over to pine forests introduced in the nineteenth century to replace oak trees which had been logged to extinction. Ligurian pine forests account for 29,000 hectares of land, now experiencing significant decline owing to climate change and pest infestation, with important landscape implications (Vivaldi, 2009). Pesto sauce is protected by a consortium and is officially listed as a Traditional Agricultural Food Product (*Prodotto Agricolo Tradizionale*) by the Italian Ministry of Agriculture.[31] A key ingredient is Genoese basil, which is safeguarded as a DOP product and is different from other types of basil owing to its small leaves and

Notes on Italy's edible landscapes: Paesaggi Pini Pinoli   33

the absence of any mint flavor. PAT pesto must have these ingredients: DOP Genoese basil, olive oil (ideally produced in Liguria), DOP *Parmigiano Reggiano*, DOP *Pecorino Fiore Sardo* (a sheep milk cheese), pine nuts (produced in the Mediterranean) and garlic. The Ligurian tradition is to use pesto with a pasta called *trenette*, cooked along with potatoes and fresh string beans. An alternative is *trofie*, a short pasta from the Ligurian town of Sori. For a wine pairing one might follow the adage of what grows together goes together and select one of the local wines produced with a rare autochthonous grape, like Pigato. Its hints of peach and Mediterranean scrub, rich palate and pleasant burnt almond aftertaste, make a wonderful pairing. Eating is an agricultural act, and to take full advantage of the occasion, one must spend time investigating all the ingredients that go into the meal, and the full range of activities required to prepare, process and distribute them. That investigation will take into consideration a complete range of factors, including consequences for the environment, biodiversity, heritage conservation, personal health and the landscape.

Environmental damage and responses

The term ecomafia was coined by the environmentalist group Legambiente in 1994.[32] This was a turning point in recognizing and documenting the worst forms of criminal action against the environment. In 2015 Law 68 was passed against environmental crimes, including environmental disaster, illegal disposal of toxic material and failure to requalify waste disposal sites. This law shows growing sensitivity to environmental needs partly as a response to the long list of ecological disasters scattered throughout the country. In the far northeast the City of Trieste in 1958 opened a 12-hectare landfill at Trebiciano, a site which was so completely uncontrolled that it became used for the disposal of all kinds of dangerous materials, including toxic industrial waste, but also anthrax and arsenic, and according to some accounts chemical weapons developed by the Italian military. The landfill was closed in 1972 amidst much public clamor, yet as of today, fifty years later, a complete regeneration has not taken place.[33]

One of the worst environmental disasters in modern European history took place at Seveso, outside of Milan, in 1976 when a failure in a Swiss-owned chemical plant released dioxin into the atmosphere contaminating hundreds of citizens.[34] About a thousand people were evacuated, an entire residential district was demolished, and the surface soil was removed for an extension of fifteen hectares. The debris generated in these emergency reclamation efforts was stored in two enormous earthquake-resistant-reinforced concrete tanks, along with the remains of 80,000 animals which died in the disaster. The region established a forest on this requalified land, called the Seveso Oak Forest. Reafforestation was an innovative strategy at the time, not only regenerating the soil but also rehabilitating the landscape. The social and political implications of environmental damage also received recognition, furnishing an opportunity for growing collective intelligence on how to treat common goods and their management (Centemeri, 2006).

In central Italy, the most significant example of an environmental disaster concerns Rome's massive landfill at Malagrotta, first opened in 1975 and then closed in 2013. At the time of its closure, it had a surface extension of 240 hectares, the largest landfill in the world. Malagrotta is on the eastern fringe of the city, in the Galeria Valley, an area whose low population density seems to have justified its use for activities with highly adverse environmental impact. Before the opening and gradual extension of this landfill, Galeria was rich in farmland and inspiring landscapes. Today it is a wasteland suspended in limbo. Though closed as a landfill in 2013, the site continues to be used for the processing of urban waste whose negative impact on the local environment reached a peak when dangerous fires broke out in its facilities in summer 2022. In ten years, little has been done to transition from a wasteland to a restored landscape. Notwithstanding considerable citizen activism, which was fundamental in determining the closure of the landfill, Galeria continues to be treated as wasteland, as seen in proposals to use another site in this same territory for a new landfill, as well as for other environmentally damaging initiatives like the creation of a plant for processing waste plastic. Running against these trends is a €250 million fund earmarked in 2022 for the regeneration of this area. This project should eventually reestablish the area's typical flora and may one day witness the replanting of historic pines. The most important citizen association of the valley is also looking for uses of the areas which can showcase its landscape potential, including urban horticulture.[35] These activists also attempt to reactivate a landscape of social relationships involving natural elements brought to life through urban gardens, sharing networks and community hubs. Yet these real-life initiatives must compete with pervasive insensitivity to the environmental contexts in a Manichaean world torn between responsibility and neglect.

The most notorious cases of environmental crimes, the ones which prompted the passage of the 2015 law, come from the south, in

Notes on Italy's edible landscapes: Paesaggi Pini Pinoli 35

particular the urban region between Naples and Caserta, identified as the 'ecocamorra' landscape, a term derived from the name of local mafia groups and 'the rise of specific illegal activities that are especially destructive of the environment' (Corona and Sciarrone, 2012:295). Until recently this area was rife with organized criminality which has no sensitivity either for environmental values nor the broader community, and certainly no concern for individual health and wellbeing. Crime groups have operated in this area since the 1960s and specialized toward the end of the last millennium in the illegal disposal of toxic waste coming in part from northern Italian industries (Armiero, 2014). This activity won for the area the epithet of the 'Land of Fires,' reversing the historic name of the 'Land of Work' (Yardley, 2014). Weak institutional control of the territory contributed to the regional waste management crisis and supported the persistence of other endemic problems such as poverty, substantial school dropout rates and political corruption. These factors are correlated with toxic waste as well as the overproduction of urban informality, with negative landscape impact. They also generate a lack of trust in institutions, hindering efforts to improve the quality of the urban environment (Berruti and Palestino, 2020).

The region of Campania is also sadly noted for the presence of Europe's most polluted river, the Sarno. The Sarno basin is located to the south of the Somma-Vesuvius volcanic complex, flanked on the south and on the east by the Apennines and bordered to the west by the Tyrrhenian Sea. The basin covers an area of 440 square kilometers with a watercourse twenty-four kilometers long. The source is in the Picentini Mountains near the city of Sarno, and the outlet is between the coastal cities of Castellammare di Stabia and Torre Annunziata, not far from Pompeii. This extraordinary water resource has many uses, including agriculture and industry. Yet spontaneous uses of the river, especially leisure activities with their beneficial impact on citizen health, have all but disappeared.

This was historically part of the Campania Felix, whose once richly productive soil is now overwhelmed by uncontrolled urban expansion. Today the area has one million residents concentrated in scattered towns with densities of up to 2,200 inhabitants per square kilometer. High population density, the excessive use of agricultural fertilizers and pesticides, along with largely unregulated industrial development, are the causes of the Sarno River's pollution. This is exacerbated by inadequate wastewater treatment (Arienzo et al., 2001). The fragmentation of territorial responsibilities is a challenge to the effective landscape management of a river running through thirty-nine municipalities, three provinces (Naples, Avellino and Salerno) and the metropolitan authority of greater Naples. Gilda Berruti and her colleagues carried out research on the Sarno River in the framework of a ministerial project on 'Landscape protection between conservation and change. Economy and beauty for sustainable development' (Berruti and Moccia, 2016). The project aimed to enhance landscape features using Kevin Lynch's image survey methodology (1960, 1990). The research project's first objective was to optimize the river's contemporary natural condition using an evolutionary resilience approach (Davoudi, 2012). The second objective was to interpret the riverscape using an inclusionary approach focusing on the relationship between the human community and the natural environment understood in functional terms as well as local perception.

A major challenge to development is the perceived distance between inhabitants and the river itself. Ordinary interaction is rare, except in isolated cases generally restricted to the springs at the river's headwaters. Local narratives show that this unpolluted section of the river is a vital landscape actively used in leisure activities. Affective ties to the headwaters contrast with the rest of the river whose weak imageability is attributable to various factors, including culverts which hide the river, and the absence of defining landmarks like bridges and historic watermills. The foul smell of polluted waters also discourages a positive image.

Flooding forces the river's attention on the inhabitants of the plain who often undertake informal efforts to reconstruct broken sections of the river's embankment. These informal activities are complemented by formal restoration actions involving multiple strategies, including environmental education as a foundation to regenerating an ecosystem bringing together people and the environment, and efforts to build a sense of the community's role in a journey aiming to attain place-focused local development (Berruti and Moccia, 2016). These formal restoration projects were started at the headwaters, with their positive perceptual connotations, and have gradually moved downstream.

Many organizations are involved in curating a new relationship with the river, foremost among which is the Friends of the Sarno Association, formally established in 2004. These associations constitute a fragment of a potential network that can further healthy development, constructing functional landscapes while promoting an aesthetics based on ecological awareness and environmental ethics (Yu, 2010).

Thanks to these efforts several hidden resources are beginning to emerge, like the remote springs of Santa Marina with its astonishing hydro-ecological landscapes, or the series of watermills and disused manor farms, traces of a productive past which can be repurposed for contemporary uses. There is also the impressive three-peaks landscape where the Sarno lets out into the Tyrrhenian Sea, with the sequence of the Capo d'Orlando promontory, Punta Campanella and the skyline of Capri. The built environment also represents an opportunity for landscape enhancement, including the multiethnic neighborhoods which stretch out along the riverbanks from Pompeii down to the coast following the via Ripuaria. Well-established yet little known archaeological sites also abound, such as the prehistoric settlement of Longola, or other sites of monumental interest such as the pilgrimage routes leading to Pompeii.

The Sarno illustrates the complex strategies involved in regenerating a fractured landscape. A significant challenge is overcoming rivalries among towns, or the excessive personal protagonism which undermines already fragile networks. One strategy used to correct these challenges is called intercropping, borrowing from the agricultural technique of cultivating two or more crops in a single allotment to exploit synergies in the lifecycles of various plants. Transposed to a landscape regeneration setting, this technique leverages the river as an infrastructure, a connector and a producer of a multifunctional landscape, strengthening synergies among the towns of the river basin.

While the protection of the environment is a supreme value enshrined in the Italian constitution, we see how this value can be neglected. Reversing trends dictated by poor institutional performance and limited environmental sensitivity is no simple matter. Yet times are changing, and all over Italy, we witness increasing interest in the stewardship of the land through such an apparently trivial activity as urban horticulture.

Notes

1 https://it.euronews.com/2022/06/27/siccita-italia-verso-lo-stato-demergenza-nazionale (Accessed July 6, 2022).
2 For the Cervi Institute information can be found at http://www.istitutocervi.it/2019/01/02/scuola-di-paesaggio-emilio-sereni-xi-ed-27–31-agosto-2019/ (Accessed May 15, 2019). For the Coraggio Cooperative consult http://www.coop-coraggio.it/borghetto-s-carlo-e-le-terre-pubbliche/ (Accessed May 15, 2019).
3 http://www.salviamoilpaesaggio.it/blog/2021/12/passante-di-bologna-cechi-dice-no/ (Accessed July 7, 2022).
4 https://www.efncp.org/policy/semi-natural-pastures-meadows/ (Accessed July 7, 2022).
5 https://www.ruminantia.it/il-pascolo-dei-domestici-e-la-conservazione-della-biodiversita/ (Accessed July 7, 2022).

38 Notes on Italy's edible landscapes: Paesaggi Pini Pinoli

6 https://eunis.eea.europa.eu/habitats/2458 (Accessed July 7, 2022).
7 I noted this variation in agricultural choices from one town to the next in my ethnography of farming communities in central Italy (Smith, 2012).
8 http://datiallefiamme.it/ (Accessed July 6, 2022).
9 https://www.provincia.vt.it/ptpg/documenti/RL_SISTEMA%20%20%20 AMBIENTALE.pdf (Accessed October 26, 2021).
10 https://www.repubblica.it/dossier/viaggi/turismo/2020/12/20/news/1_italia_e_seconda_in_europa_per_copertura_forestale_200_milioni_di_alberi_in_10_anni_per_sostenere_la_biodiversita_-278836287/ (Accessed October 26, 2021).
11 https://www.ilsole24ore.com/art/riso-produzione-varieta-italiane-aumenta-oltre-20percento-AEMPagS (Accessed October 26, 2021).
12 https://www.huffingtonpost.it/economia/2022/07/10/news/crisi_alimentare_luigi_scordamaglia_-9802264/ (Accessed July 10, 2022).
13 https://www.corriere.it/economia/consumi/22_luglio_13/grano-duro-quotazione-crolla-15percento-agricoltori-disertano-borsa-merci-de8bae9c-028b-11ed-a0cc-ad3c68cacbae.shtml (Accessed July 15, 2022).
14 https://www.comhs.org/about-us/newsroom/health-library/2022/07/12/weed-killer-glyphosate-found-in-most-americans-urine (Accessed July 15, 2022).
15 https://www.regione.piemonte.it/web/temi/agricoltura/zootecnia-pascoli-apicoltura/pascoli-lalpicoltura-piemonte (Accessed October 26, 2021).
16 https://www.aslto3.piemonte.it/wp-content/uploads/2021/03/scheda_Val_Chisone.pdf (Accessed July 11, 2022).
17 https://www.fondazioneslowfood.com/it/presidi-slow-food/razza-modicana/ (Accessed July 11, 2022).
18 https://www.colturaecultura.it/content/usi-zootecnici# (Accessed July 11, 2022).
19 http://www.ciapiemonte.it/2019/01/in-aumento-le-importazioni-di-soia-e-di-mais/ (Accessed November 16, 2021).
20 https://www.risobaraggia.it/chi-siamo/ (Accessed July 12, 2022).
21 https://www.lestagioniditalia.it/en/cappelli-variety-pasta/ (Accessed July 12, 2022).
22 http://vnr.unipg.it/habitat/cerca.do?formato=stampa&id-Segnalazione=78 (Accessed November 16, 2021).
23 Currently the City of Rome has earmarked no less than €500,000 to replace 105 trees in one of the city's largest urban parks, the Villa Borghese. These trees are now over one hundred years old, and their perilous state is considered a threat to the security of park users. https://roma.repubblica.it/cronaca/2012/05/10/news/addio_a_105_pini_storici_sostituiti_a_villa_borghese-34854801/ (Accessed June 18, 2019).
24 https://www.habitissimo.it/preventivi/potare-o-tagliare-alberi/roma/roma (Accessed June 12, 2019).
25 This orientation is all the more perplexing considering that these pine trees rarely live beyond 150 years. https://www.conifers.org/pi/Pinus_pinea.php (Accessed July 13, 2022).
26 https://agronotizie.imagelinenetwork.com/vivaismo-e-sementi/2017/09/22/pinoli-un-opportunita-per-gli-agricoltori/55399 (Accessed June 12, 2019).
27 http://www.parks.it/news/dettaglio.php?id=31351 (Accessed June 12, 2019).

Notes on Italy's edible landscapes: Paesaggi Pini Pinoli 39

28 https://www.laziogourmand.com/baccala-alla-romana/ (Accessed June 16, 2019).
29 https://blog.cookaround.com/saltainpadella/caponata-catanese-pinoli-uvetta/ (Accessed June 16, 2019).
30 https://www.ricettedellanonna.net/torta-della-nonna/ (Accessed June 16, 2019).
31 https://www.mangiareinliguria.it/ita/ricette/ricette_liguria/streppe_pesto.php (Accessed June 2, 2019).
32 https://web.camera.it/_bicamerali/leg15/commbicantimafia/documentazionetematica/27/schedabase.asp (Accessed July 13, 2022).
33 https://www.byoblu.com/2016/02/08/trieste-ambiente-rifiuti-e-malattie-la-discarica-di-trebiciano-lo-stato-risponda/ (Accessed July 13, 2022).
34 https://www.ilgiorno.it/monza-brianza/cronaca/seveso-icmesa-diossina-1.7872030 (Accessed July 13, 2022).
35 https://www.vallegalerialibera.it/2020/02/12/urban-landmark-questi-sono-i-progetti-che-vogliamo/ (Accessed July 13, 2022).

Bibliography

Artusi, Pellegrino (2003) *Science in the kitchen and the art of eating well.* Translated by Luigi Ballerini. Toronto: University of Toronto Press.

Arienzo, Michele, Paola Adamo, Maria Rosaria Bianco and Pietro Violante (2001) 'Impact of land use and urban runoff on the contamination of the Sarno river basin in the southwestern Italy,' *Water, air and soil pollution*, 131:1–4:349–66.

Armiero, Marco (2014) 'Garbage under the volcano: the waste crisis in Campania and the struggles for environmental justice,' in Marco Armiero and Lisa Sedrez (eds.) *A history of environmentalism*. London and New York: Bloomsbury, pp. 167–84.

Beaven, Lisa (2014) 'The grave of graves. The response of grand tourists to the Roman campagna,' in Gilbert Bonifas and Martine Monacelli (eds.) *Southern horrors: northern visions of the Mediterranean world*. Newcastle upon Tyne: Cambridge Scholars Publishing, pp. 79–93.

Beauregard, Robert A. (2012) 'In search of assemblages,' *Crios. critica degli ordinamenti spaziali*, 4:9–15.

Bellini, Cristina, Marta Mariotti-Lippi, Miria Mori Secci, Biancamaria Aranguren and Paola Perazzi (2008) 'Plant gathering and cultivation in prehistoric Tuscany (Italy),' *Vegetation history and archaeobotany*, 17:1:103–12.

Berruti, Gilda and Francesco D. Moccia (2016), 'Planning knowledge and process for strategies of participatory river contracts,' in Roberta Ingaramo and Angioletta Voghera (eds.) *Topics and methods for urban and landscape design. From the river to the project*. New York: Springer International Publishing, pp. 99–126.

Berruti, Gilda and Federica M. Palestino (2020) 'Contested land and blurred rights in the Land of Fires (Italy),' *International planning studies*, 25:3:277–88. https://doi.org/10.1080/13563475.2019.1584551.

Cavallo, Federica Letizia (2011) *Terre acque macchine. Geografie della bonifica in Italia tra Ottocento e Novecento.* Reggio Emilia: Diabasis.

Centemeri, Laura (2006) *Ritorno a Seveso. Il danno ambientale, il suo riconoscimento, la sua riparazione.* Milan: Mondadori.

Coldiretti (2019) *L'agricoltura in Italia in numeri.* Rome: Coldiretti.

Cole, John W. and Eric R. Wolf (1974) *Hidden frontier: ecology and ethnicity in an alpine valley.* New York: Academic Press.

Corona, Gabriella and Rocco Sciarrone (2012) 'Il paesaggio delle ecocamorre,' *Meridiana*, 73–74:13–35.

Davoudi, Simin (2012) 'Resilience: a bridging concept or a dead end?' *Planning theory and practice*, 13:299–307.

Debord, Guy (1994 [1967]) *The society of the spectacle.* Cambridge, MA: MIT Press.

De Cillis, Ugo (2004) *I frumenti siciliani.* Catania: Maimone Editore.

De Mauro, Tullio (2011) *Storia linguistica dell'Italia unita.* Bari: Laterza.

Faas, Patrick (2005). *Around the Roman table: food and feasting in ancient Rome.* Chicago, IL: University of Chicago Press.

Gentilini, Patrizia (2019) 'Prosecco patrimonio Unesco? Noi ambientalisti protestiamo e il motivo è presto detto,' Rome: *Il Fatto Quotidiano*, July 6.

Gisotti, Maria Rita (2008) *L'invenzione del paesaggio toscano. Immagine culturale e realtà fisica.* Florence: Edizioni Polistampa.

ISPRA (2022) 'Consumo di suolo, dinamiche territoriali e servizi ecosistemici.' https://www.snpambiente.it/wp-content/uploads/2022/07/EN_Sintesi_Rapporto_consumo_di_suolo_2022.pdf (Accessed August 20, 2022).

ISTAT (2021) *Coltivazioni agricole. Annata agraria 2019–2020 e previsioni 2020–2021.* Rome: ISTAT.

Kaplan, Rachel, Stephen Kaplan and Robert Ryan (2012) *With people in mind: design and management of everyday nature.* Washington, DC: Island Press.

Iannetta, Massimo (2006) *La desertificazione in Italia e il progetto RIADE.* Rome: ENEA Ente per le Nuove tecnologie l'Energia e l'Ambiente.

Latour, Bruno (2017) *Facing Gaia. Eight lectures on the new climatic regime.* Cambridge, UK: Polity Press.

Li Rosi, Giuseppe (2012) 'Terre frumentaria,' *I custodi dei semi*, April 20.

Lynch, Kevin (1960) *The image of the city.* Cambridge, MA: MIT Press.

Lynch, Kevin (1990 [1984]) 'Reconsidering *The Image of the City*,' in Tribid Banerjee and Michael Southworth (eds.) *City sense and city design.* Cambridge, MA: MIT Press, pp. 247–56.

Mutke, Sven, Javier Gordo and Luis Gil (2005) 'Variability of Mediterranean Stone pine cone production: yield loss as response to climate change,' *Agricultural and forest meteorology*, 132:3–4:263–72.

Regione Piemonte (2017) *Piano paesaggistico regionale. Schede degli ambiti di paesaggio.* Turin: Regione Piemonte.

Regione Toscana (1998) *I tipi forestali.* Florence: Edizioni Regione Toscana. http://www.regione.toscana.it/-/i-tipi-forestali (Accessed May 3, 2019).

Sereni, Emilio (2014 [1961]) *History of the Italian agricultural landscape.* Translated by R. Burr Litchfield. Princeton, NJ: Princeton University Press.

Smith, Gregory (2012) *La comunità e lo Stato. Antropologia e storia nella Marsica del Novecento.* Luco dei Marsi: Aleph Editrice.

Van Huylenbroeck, Guido, Valerie Vandermeulen, Evy Mettepenningen and Ann Verspecht (2007) 'Multifunctionality of agriculture: a review of definitions, evidence and instruments,' *Living reviews in landscape research,* 1:3:1–43.

Vivaldi, Bruno (2009) *Addio pinete di Liguria.* Turin: Informazioni Sostenibili. http://www.informazionesostenibile.info/wp-content/themes/infosostenibile2008/images/2009/10/pinete-di-liguria.pdf (Accessed May 3, 2019).

Yardley, Jim (2014) 'Mafia legacy taints the hearth in Southern Italy,' *The New York Times,* January 29.

Yu, Kongjian (2010) 'The big-foot revolution,' in Mohsen Mostafavi and Gareth Doherty (eds.) *Ecological urbanism,* Harvard University Graduate School of Design. Baden: Lars Müller Publishers, pp. 282–91.

Wine, Humphrey (1994) *Claude: the poetic landscape.* London: National Gallery Publications.

3 Urban horticulture

The COVID-19 pandemic has strengthened concern for environmentally sound practices. This concern is given voice by a growing number of citizen associations which are gaining ever more terrain. Sound practices extend from the environment to inclusionary social and economic policies, combined with efforts to promote a critical attitude to the global market. Community gardens are an eminently practical expression of these policies, linking people to nature while producing healthy food and robust communities. This activity carries forward a long national tradition of associating healthy food with local roots. The contemporary growth of community gardens is strong in urban and peri-urban settings where a return to nature is seen as a healthy alternative to the mundane commercial settings of modern food practices.

Today the consumption of locally produced food is a matter of personal choice rather than the necessity it once was. In 1938, half of the Italian population was involved in agriculture, balancing between subsistence and commercial farming. Some products were suitable for both uses, a reason why sharecroppers in central Italy in the 1950s preferred to cultivate wheat over tobacco: farmers could eat wheat (Silverman, 1963). In the Abruzzo as late as the 1970s, small-scale commercial farmers still produced most of what they consumed, leaving meagre cash reserves for non-food needs (Smith, 2012). The epic transformations of the 1950s and 1960s transfigured attitudes toward food, disembedding food from locally grounded social process. Italians moved rapidly into non-farm labor sectors, and the utilized agricultural area (UAA) began to shrink. Much later came the revolution of urban horticulture, with a new food paradigm and innovative uses of environmentally sustainable agriculture.

Cities are founded on tension between agriculture and the built environment. Rome is a case in point, with various types of agricultural activities maintaining notable visibility thanks to the city's prodigious

DOI: 10.4324/9781003279914-3

physical extension. All Italian cities contained some form of urban farming in the 1950s, with personal gardens maintained alongside commercial farm operations. Even in Venice, where arable is so scarce, urban and peri-urban farmland assured food for the city. Today this tradition has been revived in community gardens everywhere. Milan is Italy's industrial and financial heartland yet has an abundance of community gardens. In 1960, it had a hundred documented garden allotments, a number that trebled by 1980.[1] The number continues to increase, and today has reached well over 2,000 allotments scattered throughout the metropolitan territory.[2] This type of farming preserves the intimate connection between production and consumption, maintaining Italy's celebrated regional variety and seasonal focus.

The role of urban horticulture falls within a general concern present all over the world for the contemporary assessment of cities as complex ecological systems. Cities are a primary location for the management of ecosystem services: privileged environments where a proactive and targeted process of resource management can be activated and implemented. The exponential growth of urban areas has serious consequences for ecosystems which often experience deterioration owing to material and immaterial factors (United Nations, 2019). The adoption of a multifunctional approach to ecosystem services, including urban horticulture, can have significant direct and indirect consequences for sustainable human wellbeing (Costanza et al., 1997). Ecosystem services have been classed into four main groups: life support services, procurement, regulation and cultural services (Millennium Ecosystem Assessment – MEA, 2005). Urban horticulture straddles all four groups. The link between ecosystem services and community wellbeing must not be taken for granted, risking disconnection among services, urban space design and governing policies. Ecosystem disservices can aggravate the negative impact of natural events on human communities, while increasing environmental vulnerability (Lyytimäki and Sipilä, 2009; Lyytimäki, 2014). Furthermore, poor service management deepens inequalities in resource access among the social groups that constitute the urban territory (Berruti and Palestino, 2020). Urban horticulture leverages the principles of multifunctionality to enhance territorial resilience while reducing environmental risks and offering a more effective distribution of resources across the community.

Background on urban horticulture

The Food and Agriculture Organization (FAO) has advocated urban and peri-urban agriculture (UPA) since the 1990s. This policy recommendation aims to reduce the risk of making developing countries dependent on food aid from wealthy nations. Aid in foodstuff disrupts community structures, has negative impact on gender empowerment and weakens distinctive cultural heritage (FAO, 2006). Local food consumption instead boosts biodiversity and improves health, while enhancing landscape aesthetics. These benefits are defended by many citizen groups, such as *Terra Madre* which funnels resources needed to support unique local agricultural products with modest commercial value. These are the presidia which involve farmers and other territorial actors. Italy has more than 350 presidia, of which four dedicated to garlic alone (Terra Madre, 2006). Attention to distinctive food traditions enhances affective attachment to the territory, enveloping and refining personal identity. Community gardens have a similar function.

An early example of community gardens in a modern urban setting is Pingree's Potato Patch established in Detroit during the economic slump of the 1890s. In a period of rising community strife, Mayor Pingree made available 200 hectares of public land to stave off hunger among indigent urbanites, providing relief to 1,700 families.[3] Today urban farming is again playing a decisive role in the response to massive economic transformations in this midwestern city. The Urban Farming Global Food Chain initiative was kicked off in Detroit in 2005 by singer and food activist Taja Sevelle in a project which today brings together 6,000 community gardens all over the world. The aspiration is to provide food relief while building strong communities and healthy environments in every corner of the globe.[4]

Contemporary Italy, like many parts of the world, struggles with declining urban populations and the abandonment of entire city districts. The process has created the 'vacant lands' which today represent an opportunity as well as a threat. These vacant lands risk becoming accumulated wastelands comprising real landscapes of waste, or wastescapes, which need to be metabolized in their urban settings (REPAiR, 2017; Amenta and van Timmeren, 2018). Debate concerning lost land dates to the 1990s with several conflicting definitions. It can include bare soil, agriculture at the edge of urbanized areas, uncultivated land, recently razed land, derelict land with abandoned buildings and brownfields. These spaces are not only empty or unoccupied but also free and available (Gasperi et al., 2016:16). Urban horticulture is increasingly used to transform this potential wasteland into a vibrant component of healthy communities. With the 2007–8 crisis, many citizens came to see urban horticulture as more than a utopian exercise: it is a way to deal with economic hardship while strengthening critical community assets. The COVID-19 pandemic further exacerbated economic marginality while stimulating real concern for the environment. These trends boost support for a new model in governing relations among citizens and with the natural environment. This is the new food paradigm which stresses the benefits of wholesome food and sustainable food procurement. The paradigm encourages the sharing of knowledge concerning food quality, promotes social inclusion and intergenerational communication, and strengthens community integration across conventional class lines (Mangiameli, 2017).

No exhaustive catalogue exists of Italian community gardens, nor would such a catalogue have much sense given the dynamic character of these initiatives. Here we review representative cases investigated systematically starting in 2017 using the template shown in Box 3.

Box 3 Five research questions concerning urban horticulture

1 Land Tenure – What if any contractual agreement allows you access to the land? What is the cost? Do users have individual or shared rights? How much land is available for each user? What are your origins – how were you organized and who initiated the process?
2 Work – What utilities are available? What other services? What equipment is required to work the land? Is equipment owned individually or communally?
3 Management – How are production and management decisions taken, including seed choice, fertilizer and other treatments?
4 Outcomes – What are the outcomes of gardening other than food? Can products be sold? Do you have an education program for members or the public, or both?
5 Community Support and Perception – What institutional or other public support do you receive? Are any sectors of the public community hostile to your activities?
6 Long-Term Outlook, Sustainability of the Gardens – How do you see the future? Will the gardens continue with future generations? Do you have new members coming into the project to continue it?

Our investigation concerns gardens distributed across the national territory, chosen in such a way as to highlight the rich variety of experiences. The cities on which we base our study are Rome, Siena, Venice, Bologna, Turin and Palermo. We also review a special case from metropolitan Naples.

Urban horticulture in Rome

From the Middle Ages up to the modern period, Rome occupied a compact site which was but a fraction of the city's total territory. The enormous municipal boundaries were determined by Rome's historic status as home to the Holy See. At the end of the Second World War, Rome extended for 1,504 square kilometers, reduced to 1,280 square kilometers in 1992. Today Rome is one of the most extensive cities in the world, and the most extensive in Europe. Consider that New York City has a territory of 785 square kilometers, Berlin 891. Rome's population of three million residents is relatively small, determining a modest overall density of 2,159 inhabitants per square kilometer. The city's built areas are compact, leaving extensive green spaces unencumbered by human constructions. Much of the modern growth follows the consular roads leading from the city center to remote locations, creating finger-like built projections out into agricultural land or protected green spaces. This urban structure makes green spaces easily accessible to a large proportion of the population.

Rome is Europe's greenest city, although its green is notoriously neglected. The city's environmental services department is patently understaffed with only 136 municipal employees as compared to 1,200 workers twenty years ago, a far cry from the 4,500 workers considered the bear minimum to maintain this vast extension of public green (Costantini, 2017). It is a sad testimonial to weak institutional support for natural amenities. Fortunately, desolate institutional performance is counterbalanced by vibrant citizen initiatives.

City dwellers in Italy are used to having fresh and seasonal produce made available through a wide network of city-sponsored produce markets linking citizens to extensive urban and peri-urban agriculture. In the early postwar period, these markets secured produce from urban farms, or peri-urban ones, especially from Fondi to the south, or Montelibretti to the north. From these locations, farmers and stall-keepers would travel daily to the city, bringing fresh produce with them. Various factors have contributed to limiting the viability of this system of small-scale producer/retailers. Earnings from fresh produce sales are limited, and as many Italians aspire for greater monetary rewards so correspondingly produce retail in Rome has been increasingly dominated by immigrants. Lifestyle changes have made it difficult for citizens to shop daily at local markets which are only open in the mornings. Peri-urban horticulture has also had to struggle with increasing production costs, soaring prices for agricultural land and massive competition from the modern food distribution system. These factors weaken intimate city-farm linkages, now being restored thanks to urban gardens.

One of the pioneers in organizing and documenting urban gardens in Rome is an informal group called Zappata romana, a pun on *Zappatista* (suggesting guerrilla gardening) and the Italian term *zappa*, meaning hoe. These activists went public in 2010, and by 2021 documented the presence of 150 gardens in Rome.[5] In 2015, the city adopted a plan to regulate urban horticulture, issued under the progressive mayor Ignazio Marino.[6] The ordinance regulated the use of urban gardens and detailed their general aims. The chief aims were to enhance food security and safeguard the city's green spaces, rationalizing the use of natural resources. Agricultural practices were to be organic, biodynamic or permacultural, ruling out the use of genetically modified seeds and plants. The ordinance wished to enhance environmental sensitivity and create new forms of social integration. The assignment of gardens was to privilege local families in need, and the commercial exploitation of plots was prohibited. The maximum extension of an individual plot was 100 square meters, assigned for a six-year renewable period.

Urban gardens that came into being prior to 2015 were termed 'spontaneous.' These gardens were found on both public and private property, with or without authorization. The risk in allowing horticulturalists to work land without a formal agreement is that after a certain number of years the gardener can file a suit for ownership. This is called adverse possession, or *uso capione*. For a private citizen ownership loss is a form of individual privation; for a public entity adverse possession is instead detrimental to the entire community since land is no longer available for social uses. No public record is kept of the incidence of adverse possession in Rome, but one often encounters such cases, as will be seen below.

Of the many gardens documented by Zappata romana two stand out, and not only because they are Rome's first experiments in contemporary urban horticulture. One case is the Consolata urban gardens on the western side of the city, the other the Aniene gardens on the eastern fringe.

The Consolata gardens were established on an extension of the lost space which abounds in the Roman periphery, a water-logged depression left free in the midst of intense construction concentrated on the surrounding elevations. This entire area had been farmland from the late nineteenth century, generating produce for sale at the local urban markets. Then farmland gave way to residential neighborhoods. The elevations were built up as compact though often unauthorized and unplanned communities, and the depression they enclosed was left as a residual unoccupied space because it was subject to heavy flooding. When in the late part of the twentieth century the valley was equipped with storm drains the construction of a new residential district was no longer an option since the defense of zoning regulations had become stringent. The valley became lost space. It was in these circumstances that local citizens began to create spontaneous gardens in the residual depression. The process started in the 1970s as instances of guerrilla farming with gardens squatted on publicly owned land. The municipal police were inconsistent in handling these violations, although they would periodically issue fines for illegal occupation. The inconsistency of their actions is evident in the fact that one guerrilla gardener on a remote inner plot pressed a claim for adverse ownership which ultimate legalized his claim to a plot of 200 square meters. Today the owner has fenced in his garden but has no access; he reaches it thanks to informal easement rights accorded by gardeners who are now authorized to work the public land on his southwestern border.

Starting in the 1970s, the spontaneous gardeners made every effort to secure a rental agreement with the city of Rome, but the municipal authorities were unresponsive. In 1979, the municipal police stepped up their efforts to control the situation, issuing fines and eviction orders. But their efforts were intermittent, creating an ambiguous situation that carried up to 2011 when the city finally authorized what became Rome's first formal community garden. Their management was entrusted to a legally recognized association called Fossa Bravetta (*fosso* means ditch or depression, Bravetta is the place name). To everyone's surprise, the agreement was reached thanks to the intervention of Rome's unprecedented far-right mayor. This mayor was widely considered to be supported by the city's powerful lobby of real estate developers, notoriously hostile to community gardens because they deprive developers of an investment opportunity. Today's gardeners surmise that gardening rights were granted as a pretext to award a lucrative contract to the developers who supported the mayor. Indeed, no garden in Rome has such impressive facilities, from a robust fenced enclosure of stone and iron bars, to a brick club house with bathrooms and kitchen facilities, to the artisan well with water distribution reaching each plot, to the hedges which surround individual gardens, and the generous work sheds. The mayor authorized a budget of half a million euros to develop twenty-six garden plots, at the cost of eighty euros per square meter. Other gardens in Rome are developed for a tenth of that cost. The suspicion of malfeasance was strengthened when the mayor was convicted of graft and corruption in an investigation which the press dubbed 'Rome the Capital of Mafia.' Yet aside from questionable motivations, the project is extremely successful. Only about a third of the fenced in area managed by the Fosso Bravetta Association is occupied by gardens. The rest is a park with benches, and play spaces for children, all well maintained by the association. The liberality of the mayor's action was such that the association is required neither to pay rent nor utilities. No other garden in the city enjoys such privilege.

In close proximity to these gardens, another forty plots were developed subsequently. By 2019, there were sixty-five gardeners in the valley, all from the surrounding neighborhoods, mostly seniors, all of Italian origin, and all but five men. All the plots are well maintained and the community that works them is well integrated. Their highly visible activities have triggered broad interest in community gardens: today the association has a waiting list of 180 citizens.

A contrasting example of urban gardens is found in the Aniene Regional Park on Rome's eastern periphery. This park was established in 1997 along the banks of Aniene River, an important tributary of the Tiber. The park falls under the authority of Lazio Region. Not far from the authorized gardens, within the boundaries of the regional park, one finds squatted horticultural plots which sprang up as liberty gardens around the time of the Second World War. Today these plots allegedly pass hands through illegal cash sales. They are fenced in with informal materials, including the framed mesh normally used to support the mattress of a bed. The plots have no water access, and many gardeners use bulky plastic tanks to store water which is trucked in. The local police patrol the area but are under orders not to interfere with the gardening activities. According to local accounts, one gardener, who acquired his plot through an informal cash transaction, attempted to build a stone house on his plot. The municipal police intervened and stopped the building, although without depriving the gardener of his patch.

Outside the liberty gardens, the regional park authorities and the municipal police make a concerted effort to regulate all forms of land use. The land is owned by the City of Rome but falls under the authority of Lazio Region which in turn assigns management to a public agency known as Resources for Rome.[7] Resources for Rome have in turn entrusted day-to-day management of a significant portion of the park located along the Aniene's northern bank to a citizen association called Together for Aniene. The association's stone club house was built at the expense of Resources for Rome. The community gardens falling under their authority were first established in 2011 and subsequently expanded using EU funds managed by Resources for Rome. Together for Aniene was founded in 1999 by citizens from the adjoining neighborhoods as an environmentalist group aiming to promote and protect the parkland. They started by offering courses and activities involving the environment. One of their courses is on wildflower harvesting, teaching citizens how to gather spontaneous edible plants in the park's open fields.[8] The presence of the association has had important impact on the landscape. It is enough to compare the northern bank of the Aniene, where the association is located, to the southern bank. The southern bank was home to entire shanty communities in the early postwar years and throughout the 1950s. Today the shanties have been dismantled, but many claims of adverse possession have been made, leaving a scattering of constructions in all different styles, some commercial concerns some used as homes. Fortunately, today the park authority effectively obstructs unauthorized land use. But the consolidated presence of low-quality buildings dating to a period when controls were weak diminishes landscape value and limits opportunities to enjoy the area's natural qualities. In hindsight, it is easy to see that allowing citizens to have regulated access to green areas which the city is unable to protect owing to a lack of resources is a way to ensure a high standard of environmental control.

Together for Aniene developed gardens on the northern bank of the river in two different phases, reaching a total extension of ten hectares. The first set of gardens was developed at the gardeners' personal expense, including initial plowing and leveling, the creation of water access points, shared tool sheds and basic fencing. The creation of a second set of gardens was made possible, as noted, thanks to EU funds. Both sets have rudimentary fixtures, like the cane fences that define rather than enclose spaces. The water access points are deliberately limited in number, requiring that a single source be shared among different allotments, encouraging gardeners to collaborate. The second set of gardens developed at public expense required an investment of 40,000 euros, equivalent to 8 euros per square meter. The first set of gardens were assigned in 2011 on a first come first serve basis, while the second set were assigned in 2015 using the new assignment criteria established by the city ordinance issued that same year. Both sets of assignments are regulated by a contract signed between the association and the local municipal authority. According to the ordinance, plots cannot be sold and when a holder gives up a plot it is returned to the association for reassignment. No chemical fertilizers or pesticides are allowed. Crop rotation is the rule, alternating between nitrogen fixing plants likes peas, chickpeas and beans, and plants that absorb nitrogen. Minimal rent is paid to the association, 180 euros a year for a fifty-square-meter plot. Some plots are worked by more than one gardener, with as many as four or five participants per allotment.

Urban horticulture in Siena

Siena is an excellent example of how effective management of the territory can maintain a balance between nature and the built environment. Siena has a compact built fabric, yet historical maps show that it has long defended a mosaic of green spaces some of which still provide inhabitants with fresh local produce.[9] The defense of these green spaces was also mandated in the 1956 master plan. The most substantial of these spaces is an open field of about ten hectares extending from the central Piazza del Mercato down to the city's historic walls. This open field, contained within the fortified valley, was once a residential district then decimated by the great plague of 1348. It was subsequently abandoned, and by the sixteenth century had become an extensive garden known by its current name, Orto de' Pecci.[10] Today the valley falls under the stewardship of a type B social cooperative known by the name of La Proposta, founded in 1983 by a group of young people who worked in the adjoining psychiatric hospital. La Proposta started as a group of gardeners, and gradually expanded their activities to include a restaurant. To qualify as a type B social cooperative 40% of their workers must have disabilities. These are sometimes termed fragile workers, a condition which includes citizens with documented disabilities that impair their ability to work, citizens with mental health issues, those with a history of substance abuse, convicts on work release, and other cases as well.

Prior to the 1980s, the Orto de' Pecci was worked by residents of the psychiatric hospital as a therapeutic exercise and as a source of healthy food. Then in 1978 psychiatric clinics all over Italy were closed, and the mission of the garden was redefined to serve the needs of a broader community. The social cooperative took over the gardening activities, and later extended their services to schoolchildren who could have meals using locally grown produce while learning about urban horticulture. They still provide this service but have expanded and now maintain a successful restaurant adjoining the gardens, keep up their gardening and offer educational services dealing with the environment. They also hire out landscaping expertise, helping citizens create attractive and edible landscapes in the broader urban territory. The cooperative today employs more than fifty workers.

Many areas of Italy use horticulture as part of a rehabilitation strategy for citizens with different forms of fragility. The most famous case is San Patrignano on the Adriatic coast which hosts a community of 800 residents with a history of substance abuse. They have an extraordinary rehabilitation success rate. The use of horticultural therapy does not stop with substance abuse. In Rome, for instance, the Catholic relief agency Caritas maintains a facility for juveniles who have broken the law and are required to reside in a correctional facility. They use horticulture to pursue their rehabilitative aims. Horticultural projects are also found in the juvenile correctional facility of Nisida in Naples, as well as in the Secondigliano prison in a project coordinated by the social cooperative Man and Wood.

Siena is also home to more conventional urban horticulture projects, notably the gardens established in the public housing complex of San Miniato, a satellite community about ten kilometers from the historic city. The provision of public housing was mandated all over Italy by Law 167 of 1962. One of the law's requirements was that districts with public housing furnish ample services to residents. Often these districts contain facilities for private and public firms, as a source of local employment. This is the case of San Miniato. The district's core contains public housing, mixed in with clusters of middle-class apartment buildings, a branch of the university, offices owned by the Bank of Siena and student housing. As is often the case with public housing projects, funding has been discontinuous, and after many development interventions, a substantial lost space was left near the central public housing unit. The vocation of this space was still unclear in the early millennial years, until in 2015 it was decided to establish community gardens there. This was part of a project launched by Tuscany Region under the heading 'One Hundred Thousand Gardens in Tuscany.'[11] The gardens satisfied two planning needs, one was to integrate a neighborhood comprised of a diverse range of citizens, achieved through gardening as well the organization of meetings, lectures, and other community events. The other need was to complete the physical design of this community, harmonizing it with the natural landscape surrounding Siena. Low-cost urban gardens blend in with the hills, complementing the curvilinear structure of the public housing unit. The gardens resonate with the natural environment while providing healthy food for the community.

Urban horticulture in Venice

Gardening in Siena with its strong public support contrasts with the situation in Venice. The urban setting is much different, as is the political context which has generally favored a neoliberal approach to shaping urban growth as compared to the planning focus found in Tuscany. Venice struggles with potent environmental challenges, exacerbated by modern mass tourism and the dramatic decline of its working-class population. In 2021, the region passed a €4 billion plan called Venice World Sustainability Capital. It is an ambitious program involving sustainable tourism, courses on environmental education, sustainable energy and affordable housing. The plan does not mention community gardens, which in fact have received limited public support in this island city.

Historically, produce consumed in Venice was grown in the lagoon in a commercial activity which declined after the 1950s, and eventually disappeared. With small plots and high transportation costs, local producers could not compete with the modern food system. Yet in the lagoon today there has been a modest return to farming, supported in part by the local market for regional specialty products, in part thanks to urban gardens made available by the city authorities. Market-oriented urban farming has enjoyed increasing success largely thanks to initiatives undertaken by the well-known environmental activist Michele Savorgnano, and his Diffuse Urban Farm, known in Italian with the acronym FUD.[12] This remarkable project makes the products of urban farming available to local restaurants, and also organizes other initiatives supporting local roots and identity, including training, meetings, lectures on environmental topics and theater events. FUD gardens are concentrated mostly in the former industrial and working-class district of the city, Giudecca Island. FUD revives gardens that were once scattered throughout Venice, housed within convents or other communities which consumed the produce grown in these spaces. FUD has revitalized these areas through agreements with private and public entities which give them access to fragmented urban farmland. The most important of these entities is the Institute for Shelter and Education (IRE), a former religious association now a municipal agency specialized in welfare services. FUD is part of a network containing other like-minded associations such as Green Spaces and The Lagoon in a Glass. They coordinate efforts to revive abandoned urban gardens. The gardens worked by FUD and their associates produce some 300 varieties of edible plants, many of them local specialties. They produce thirty five varieties of tomato, fifteen types of basil, twenty varieties of mint, as well as vines and fruit trees from apricots to plums. The catalog of edible products goes on with flowers like poppy, marigolds, hibiscus, nasturtium and herbs like sage, estragon and fennel. They also produce celery, beets, artichokes, garlic and red onions. They use exclusively organic farming, and where possible permaculture, techniques now adopted by other Venetian urban farmers with the aid of training programs delivered by FUD.[13] In the absence of any form of public support, their activities rely wholly on sales to the local market. Some clients are restaurants, including the high-end Cipriani Restaurant whose chef makes daily purchases directly from the gardens. These restaurants serve the needs of Venice's abundant tourist flow while helping maintain sustainable local development at no public cost.

Before FUD came into being another urban farming initiative found a home in the woman's prison of Giudecca. This correctional facility was once a convent encircling an extensive open space used to produce food for the nuns. The garden is substantial, equivalent to 6,000 square meters, or half a hectare, including a greenhouse which allows convicts to produce food all year long. Their products are sold through a local market with stalls operating every Thursday morning right outside the prison doors. Local citizens enjoy the bounty of the native soil while producing revenue for prisoners in an activity which also has a rehabilitative function. Since 1994, the distribution of these products has been coordinated by the Rio Terà dei Pensieri Cooperative, and what is not sold through local retail outlets is sold through local food networks. The initiative also appeals to the local tourist market, using prison-grown aromatic and medicinal plants to create cosmetic products sold through a hotel network.

Community gardens are also available to residents both on the mainland and in the historic center. Our investigation focused attention on the city center given its unique character within the Italian landscape. The greater part of the city's population in fact lives on the mainland, where five of the city's seven horticultural gardens are located. The seven locations today make available a total of 340 allotments.[14] When we visited the two gardens in historic Venice and Lido Island in 2019, almost all the gardeners appeared to be elder Italian males. The situation may be changing, and the municipal office responsible for the gardens claims that 60% of new requests for allotments come from women.

Lido is an island with easy ferryboat access to the rest of the city. In 2019, it had 120 garden plots. The only criterion for assignment of a plot is that the user be a city resident and at least sixty years old. In 2019, the average gardener was seventy-four years old. The other set of horticultural allotments are in Dorsoduro, one of the historical center's six districts called *sestieri*, a former warehouse district running along the Giudecca canal. It is central but off the main tourist routes. Here there are 40 plots each of 20–25 square meters, located in the former convent of Angelo Raffaele. The city owns both sets of allotments and has managed the gardens directly since 2017. Previously gardens were managed by the borough, each with its own set of rules, and none with sufficient resources to support the activity. In 2017, the gardens passed to the city which now implements uniform rules and the promise of more substantial support. Requests for allotments today must be accompanied by an income statement and priority is given to residents in need. Yet, there seems to be limited competition for allotments and in both locations, some plots had been abandoned by the spring of 2019. In Dorsoduro, only 30 of the 40 plots were occupied, and in Lido only 107 out of 120. Urban farmers at the time claimed that the gardens receive insufficient maintenance and for that reason gardeners give up. The Lido allotments in particular showed signs of neglect, with several overgrown plots.

A city report produced in 2017 (Marotta et al., 2017) catalogued the problems faced by gardens, including abandon, poor maintenance and rodent infestation. Since that year management of the gardens has been coordinated by the Garden Project Team of Venice's Social Cohesion Department whose activities mostly concern the city's elderly population. While unified management of urban gardens suggested stronger support, in 2019 the conservative mayor proposed to convert part of the Dorsoduro gardens into tennis courts. This transformation would serve the needs of affluent residents to the detriment of gardeners who are mostly working class, suggesting continuity with the forces which have driven the gentrification of Venice's center over the past half a century. We will see if the Venice World Sustainability Capital proposal will open the way to more inclusionary and garden friendly policies.

Urban horticulture in Bologna

A recent study shows how deeply established urban horticulture is in Bologna (Sanyé-Mengual, 2018). Bologna is the capital of Italy's second most important region in terms of agricultural revenue, and one of its most industrialized. These productive activities generate substantial wealth, but also high rates of pollution. Bologna is one of Italy's most progressive cities, with a long tradition of left-wing governments. It also has a high percentage of immigrants: 15% of the population as compared to a national average of 8.5%. As in many parts of Italy, its residents rely heavily on commercial food products for their ordinary needs but express a preference for urban food products representing distinctive social and cultural values, proximity and quality.

The residents are willing to pay premium prices for these products. Bologna has the highest rates of urban agriculture in Italy, with 2,700 gardens available to a population of some 400,000 citizens. This makes for one garden plot per 140 residents, as compared to the one per 750 of Venice. The Bologna gardens represent variegated approaches to ownership and management, from municipally supported allotments to food coops, to guerrilla gardening on abandoned flowerbeds. The city boasts Italy's first experiment in rooftop gardening implemented in a public housing project. Gardens have many functions, including strengthening food security for families in need and the requalification of abandoned industrial areas. Urban agriculture occupies an area of twenty-nine hectares, including urban gardens (sixteen hectares) and other garden typologies (private gardens, gardens in monasteries, schools, squatted gardens) (thirteen hectares). Bologna urban gardeners are mainly male (69%), Italian citizens (94%) and above sixty years of age (77%). But new trends are emerging and in the waiting list of more than 4,600 applicants, 43% are younger than forty years old and almost half are women. As compared to other regions, Bologna also stands out for the number of foreign citizens requesting gardens, now 10% of the total.

Bologna is well known for its compact layout and excellent public housing. A key strategy has been to structure growth to the north, limiting encroachments on the green hills to the south. But even in the north keen attention is paid to landscape quality. An effort to build a new beltway through the northern area met with strong opposition by one Italy's most esteemed urban planners, Pier Luigi Cervellati.[15] He helped contain urban sprawl in the 1960s and does not intend to stand by idly when witnessing new challenges to such a well-curated balance between the built and the natural environment.

The earliest authorized community gardens were established in proximity of public housing. This bolsters food security and helps showcase excellence in inclusionary urban design, putting Bologna's public housing on the vanguard of contemporary urban trends. The first gardens were built in Pilastro, a neighborhood founded in 1966 to accommodate the heavy influx of immigrants from other parts of Italy drawn to the promise of employment in industries not far from this location. It has a high concentration of public housing, the most impressive component being the 700-meter-long building known as the 'big comma' owing to its curvilinear design. It was built in 1976 using funds made available by Law 167 of 1962. Bologna's first official urban garden was established adjacent to the big comma by an association called the Circolo La Fattoria, meaning literally the farm club, managing a farm that was established in 1980. Circolo La Fattoria is a community center sponsored in 1967 by what was then the communist party's recreational group (ARCI). Bologna is colloquially termed 'the red' owing to its historic loyalty to the Italian Communist Party (PCI). ARCI worked in close coordination with the PCI to ensure the provision of recreational services in even the most disadvantaged parts of the city.[16] La Fattoria is a remarkable community resource, maintaining community gardens, an animal farm producing cheese and an active community center. It offers a rich calendar of workshops on such topics as biodiversity, seasonality and sustainability, as well as literacy courses for foreigners, and dance classes. A total of 120 garden plots are located on publicly owned property adjacent to the housing project. Eighty gardeners work these plots, assisted by two asylum seekers who are paid employees of the association.

The largest community garden in Bologna is also in Pilastro, operated by the Association of the Via Salgari Gardens founded in 1980. The association manages 420 garden plots of some forty square meters each, worked by 500 association members. The land is publicly owned, and the gardens are managed communally by the plot holders, coordinated by an elected president and a board of directors. As with La Fattoria, gardening is one of many activities, and occupies only 60% of their 30,000-square-meter property. The rest of the space is a park for residents, well maintained by association members.[17] The site has a clubhouse with assembly areas, kitchen facilities and bathrooms. The success of gardening is seen in a waiting list of 200 applicants. An inclusionary policy mandates that 10% of the plots be reserved for immigrants. Plot assignment is determined by a public competition based on age, income and proximity to the garden. The rental fee is 45€ a year. The cultivation technique is strictly organic.

The first rooftop garden established on a public housing project is located in the district of San Donato, another working-class area with a high concentration of public housing right outside the railroad tracks defining the northern limits of the city's historic growth. This is a densely populated area with limited open space. The garden was created in 2010 in collaboration between the municipal authorities, the University of Bologna, and the Association BiodiverCity. It occupies 250 square meters, producing healthy food while promoting community cohesion.[18]

Urban horticulture in Turin

Turin is Italy's most polluted city deriving from the heavily industrialized Po River Valley, as well as numerous specific contaminated sites left over from its long industrial history.[19] Concerted efforts over the past twenty years have allowed Turin to remediate the situation significantly. Turin was Italy's Detroit, and is making even more progress than its American twin in progressing from Motown to Growtown. The process began in 1993 when the new mayor forged a network of new actors, moving away from the logic of a one-company-town gravitating around car producer Fiat, to a polycentric town with strong attention to the environment (Vanolo, 2015). The showcase for Turin's transformation is the former industrial district known as Spina 3, an area which once employed 20,000 blue collar workers on an area occupying one million square meters. It was targeted for redevelopment in the 1995 master plan at the cost of one billion euros.

Community gardens have played a significant role in this process of environmental remediation. One of the best-known gardens is the Hortus Conclusus, situated in the repurposed Spina 3 industrial park within the roofless walls of a former Fiat manufacturing plant. This indoor-outdoor garden has potent symbolic value notwithstanding its diminutive size, comprising some twenty plots worked by forty gardeners selected through a competitive process involving age, income and place of residence. The land passed from private hands to public ownership in 2000 and required extensive remediation before it could be given over to gardening. The process lasted ten years. Today the Hortus Conclusus is managed by an association called Innesto (meaning 'graft' and suggesting new life for exhausted plants). Innesto not only manages the gardens, it promotes multiple initiatives revolving around the concept of active citizenship, including workshops on environmental themes, poetry readings in the gardens, afternoon drinks and light meals. All these initiatives encourage community cohesion while leveraging the theme of urban regeneration through urban gardening.

Innesto is part of another network called Orme (meaning 'footsteps' and suggesting a shared journey) bringing together multiple urban gardening associations. Orme is supported by the San Paolo Foundation and the Food Atlas of Metropolitan Turin. The Atlas has coordinated the expertise of various universities to produce an extraordinarily detailed survey and assessment of the metropolitan food system starting at the neighborhood level. It is Italy's most sophisticated system of mapping out urban food requirements. Orme started their activities with a meeting held in 2016 at Turin's Home of the Environment. By 2021, Orme listed twenty-four community gardens representing an array of management approaches. The French hobby story Leroy Marlin in 2017 launched a project called Do It Yourself Gardens developed on its own property for use by schoolchildren. By 2021, they had involved

350 youngsters in the project. The German food chain Lidl made available the rooftop of one of its megastores for a 1,400-square-meter community garden. It is managed by an international association called RE.TE whose local aim is to serve disadvantaged citizen through the provision of urban gardens at no cost. Another rooftop garden, Orto Alto Ozanam, is located atop a former steel foundry now owned by the City of Turin. It has meeting places and coffee shops as well as gardens. These are only a few examples of multiple local gardening initiatives, the most sophisticated range to be found anywhere in the country. Following the pattern established in the 1990s networks play a key role, bringing together foundations, universities, public entities and myriad citizen associations.

Urban horticulture in Palermo

Palermo is Italy's fifth largest city, and the one hardest hit by mafia-type criminality which has had devastating impact on so many areas of local life, as well as on the fabric of the city. Restructuring efforts started with the Sicilian Spring in the late 1980s, consolidated in the urban rehabilitation plan of 1993, drawn up by one of Italy's most distinguished urban planners, Pier Luigi Cervellati (Schneider and Schneider, 2003). His plan consisted of identifying regeneration nuclei in the decaying urban center, especially the Kalsa district. The success of his plan has been enormous and has driven a virtuous process of investment which has transformed a major portion of the historic center. But much needs still to be done to regenerate the city, and part of the process could be driven by urban horticulture.

One of the oldest urban horticultural experiments is called Palermo Urban Gardens. They operate to the north of the historic center in an area characterized by the relative abundance of green space. A thirty-square-meter allotment costs €25 a month, or about €300 a year. This is a much higher price than paid in Bologna or Rome, possibly because community gardens in Palermo seem all to be established on private property. Indeed, being a private initiative means there is no selection process based on age or income. The City of Palermo furnishes little support for community gardens, yet some initiatives are quite ambitious. The Orto Capovolto project stands out in particular, situated in the dense central district of the city called Ballarò. The garden is supported by community associations and local schools and produces food in garden boxes erected on the pavement. It has an important educational function and helps promote community integration through gardening and street food festivals.

Palermo disposes of an abundance of property confiscated from convicted mafia criminals. Some of these properties are now farms worked as single units by associations specialized in furnishing support to citizens who are in a state of significant need. None are assigned to associations representing individual gardeners. This is the case of confiscated property in the district of Ciaculli, home to Sicily's most important mafia groups in the early postwar years. In 2015, the city assigned a sizeable plot of agricultural land to Green Cross, a mutual aid association with activities all over the country providing support for needy families.[20] This assignment was part of a project involving several pieces of confiscated property, all awarded to associations serving the needs of indigent citizens. Sets of allotments provided to individual gardeners are instead established on private land, and even the gardens organized in proximity of Palermo's most important public housing project, ZEN, are operated on private soil with no public support. A single plot rents for €300 a year.

In this connection, we must note the strong media attention directed at an unusual urban activist, Biagio Conte. Brother Biagio, as he is called locally, comes from a family of wealthy entrepreneurs, and abandoned all the comforts of his privilege to assist Palermo's substantial poor population. Over the years he has assembled a wide array of resources, coordinated by the Mission of Hope and Charity which provides shelter to 1,200 indigent citizens. Among its assets is an important historic residence, Villa Florio, made available at no cost by the Archdiocese of Palermo in 2017. It shelters fifty homeless citizens who also cultivate organic produce on the adjoining farmland. Fifteen members of the community are immigrants.[21]

In 2018, the Archdiocese of Monreale made available to Brother Biagio's community fifty hectares of farmland given over to wheat cultivation in the Belice Valley to the south of the city. A few years earlier investigating authorities had discovered that mafia criminals tied to Italy's most ferocious killers had a hand in working these lands. When the Archdiocese learned of this circumstance, they cut off all ties with the former management and entrusted the lands to Brother Biagio.[22] This is an effective way to diminish returns to mafia while helping citizens in need.

Urban horticulture in metropolitan Naples

In some cases, urban horticulture is practiced in marginal areas beyond the public limelight, beyond policy scrutiny, areas of innovation where space regeneration, the return to nature and cross-sectoral governance go hand in hand. These are often wastelands discarded in a distorted process of urban metabolism, areas which must be reutilized in the city's growth processes. Reclaiming wastelands restores an urban ecosystem based on an appropriate cycle of urban metabolism while allowing the community actively to regain common goods lost in the city's growth processes. An example of the dual effort to restore urban ecosystems while regaining common goods is that of property confiscated from the mafia in the Neapolitan hinterland, and then assigned to the Antonio Esposito Ferraioli Farm. The Ferraioli Farm is located in a strategic position in the metropolitan area of Naples, in the town of Afragola, near the high-speed rail station in the middle of a peri-urban area with dense road infrastructures, waste infrastructures, abandoned land, illegal constructions and fragments of countryside. It is the largest confiscated property in metropolitan Naples, with twelve hectares of agricultural land, mostly orchards, and an 830-square-meter derelict farmhouse.

Once owned by a camorra boss, the farm was expropriated using legal provisions we will explore further in the next chapter, and then assigned to a group of associations and social cooperatives which chose to dedicate their activities to the memory of a young trade union activist who was murdered by local crime groups in 1978. The path to re-appropriating this common good started in July 2014 and was completed in March 2017 when the city of Afragola assigned the property at no cost using a ten-year renewable contract. The property had been expropriated in 1982, and in the intervening period was used as an illegal waste site as well as informally for the commercial production of peaches. A few days after assignment, the antimafia group planted 400 apple trees using a local variety, trees which were promptly stolen. This act of intimidation provoked a powerful community response, when the organizers called an 'Anti-Camorra Open Day' which attracted a public of 1,500 citizens to replant the stolen trees. In December of the same year, the associations decided to dedicate one hectare of land to vegetable gardens, in response to a need expressed by many citizens coming from the greater metropolitan area where community gardens are rare. Initially one hundred gardens of 100 square meters each were assigned to citizens coming from the broader metropolitan area. Subsequently some gardens were divided, and by 2022, there were 308 gardens ranging in size from fifty to one hundred square meters. Today there is a long waiting list for people hoping to join this farming community.

Garden plots are assigned to individual citizens for no rent using a one-year renewable contract. Garden management is directed by an assembly of gardeners which meets every third Saturday to decide changes in the gardening regulations. The planting of fruit trees is one of several activities shared by all the associations working together, and by 2022 1,700 fruit trees had been planted using varieties typical of the territory to create a living museum of biodiversity.

The associations which take responsibility for managing the farm stress the importance of landscape regeneration, environmental education, and opportunities to build strong community bonds while working the land according to sustainable standards. They are also interested in creating jobs, and part of the land is used by employees of the associations to grow produce which is sold to the public in informal structures built on the property for this purpose. The Ferraioli Farm is unusual from the standpoint of the urban horticultural practices we considered in the previous pages. To start with, it is a confiscated mafia estate. Furthermore, it draws citizens together from a wide metropolitan area rather than representing an example of community horticulture, which is quite rare in Naples. Finally, it combines gainful and non-gainful activities. The non-gainful activities are the horticultural plots assigned free of charge. The gainful activities instead concern the orchards which are maintained by professional farmers employed by the associations, and an extensive organic farm used to grow produce for local sale. These gainful activities generate funds which help the associations develop the property (Berruti, 2019). In the next chapter, we will understand better the activities of associations working confiscated mafia estates.

Notes

1 https://thesubmarine.it/2016/06/10/come-gli-orti-urbani-hanno-ridisegnato-milano/ (Accessed November 25, 2021).
2 https://www.lacucinaitaliana.it/lifestyle/ambiente/orto-urbani-a-milano/?refresh_ce= (Accessed November 25, 2021).
3 https://www.hourdetroit.com/community/pingrees-patches/ (Accessed November 27, 2021).
4 https://www.urbanfarming.org/welcome.html (Accessed November 27, 2021).
5 http://www.zappataromana.net/en/mappa/ (Accessed June 15, 2021).
6 City of Rome ordinance number 38 of July 17, 2015.
7 We conducted various interviews in the spring of 2019 with Lucio Zaccarelli, the secretary of the association which manages the gardens.
8 https://www.aniene.it/onlus-roma/insieme-per-aniene (Accessed July 7, 2021).
9 http://www.ortodepecci.it/webnew/la-proposta/la-storia/ (Accessed June 15, 2021).

Urban horticulture 63

10 http://www.ortodepecci.it/webnew/a-tavola-con-la-storia-il-catasto-dellorto-de-pecci-1318–1330/#more-1520 (Accessed June 15, 2021).
11 https://www.regione.toscana.it/-/l-iniziativa-100mila-orti-in-toscana (Accessed April 21, 2019).
12 https://www.fud.email/post/2015/02/11/s-erasmo-si-parte-con-la-progettazione-e-i-primi-lavori (Accessed April 11, 2019).
13 https://www.greenme.it/viaggiare/italia/gli-orti-nascosti-di-venezia/ (Accessed April 11, 2019).
14 https://www.genteveneta.it/mestre/gli-orti-urbani-piacciono-alle-donne-triplicate-nel-comune-di-venezia/ (Accessed July 20, 2021).
15 https://corrieredibologna.corriere.it/bologna/notizie/cronaca/2016/6-settembre-2016/passante-comitati-no-si-blocchi-tutto-240872378324.shtml (Accessed December 2, 2021).
16 Circolo La Fattoria. 2015. http://www.circolofattoria.it/chi-siamo/ (Accessed February 22, 2019).
17 *Bologna città degli orti. Orticoltura urbana tra tradizione e nuove tendenze.* Bologna: Fondazione Villa Chigi. http://www.comune.bologna.it/media/files/relazione_orti_urbani_a_bologna.pdf (Accessed February 25, 2019).
18 https://una.city/nbs/bologna/community-garden-gandusio (Accessed December 2, 2021).
19 This is specifically in terms of the number of heavily polluted sites found within the regional territory which have been classified, mapped out and earmarked for remediation starting in 1997. Piedmont once was among the most polluted regions in Italy, but has managed to reduce the polluted sites (SIN) to a small number thanks to many effective environmental remediation programs which have intensified in recent years. https://www.mite.gov.it/sites/default/files/bonifiche/contenuti/Anagrafe_07092015.pdf (Accessed December 2, 2021).
20 https://www.comune.palermo.it/palermo-informa-dettaglio.php?tp=1&id=8418 (Accessed November 12, 2021).
21 https://www.redattoresociale.it/article/notiziario/la_comunita_dove_senza_dimora_e_migranti_vivono_insieme_e_coltivano_la_terra (Accessed November 12, 2021).
22 https://www.monrealenews.it/cronaca-monreale/cronaca-varia/21382-la-gestione-dei-terreni-del-santuario-di-tagliavia-alla-missione-%E2%80%9Csperanza-e-carit%C3%A0%E2%80%9D-di-biagio-conte.html (Accessed November 12, 2021).

Bibliography

Amenta, Libera and Arjan van Timmeren (2018) 'Beyond wastescapes: towards circular landscapes. Addressing the spatial dimension of circularity through the regeneration of wastescapes,' *Sustainability*, 10:12:4740. https://doi.org/10.3390/su10124740.

Berruti, Gilda (2019) *Fuori norma. Percorsi e ragionamenti su urbanistica e informale.* Rome: INU.

Berruti, Gilda and Maria Federica Palestino (2020) 'Contested land and blurred rights in the Land of Fires (Italy),' *International planning studies*, 25:3:277–88. https://doi.org/10.1080/13563475.2019.1584551.

Costantini, Valeria (2017) 'Hanno ucciso il verde con la strage di giardinieri,' Milan: *Corriere della Sera*, January 17.

Costanza, Robert, Ralph d'Arge, Rudolf de Groot, Stephen Farber, Monica Grasso, Bruce Hannon, Karin Limburg, Shahid Naeem, Robert V. O'Neill, Jose Paruelo, Robert G. Raskin, Paul Sutton and Marjan van den Belt (1997) 'The value of the world's ecosystem services and natural capital,' *Nature*, 387:253–60. https://doi.org/10.1038/387253a0.

FAO (2006) 'Economic controversies over food aid,' in *The state of food and agriculture 2006. Food aid for food security?* Rome: Food and Agriculture Organization.

Gasperi, Daniela, Giuseppina Pennisi, Niccolò Rizzati, Francesca Magrefi, Giovanni Bazzocchi, Umberto Mezzocapo, Monique Centrone Stefani, Esther Sanyé-Mengual, Francesco Orsini and Giorgio Gianquinto (2016) 'Towards regenerated and productive vacant areas through urban horticulture: lessons from Bologna, Italy,' *Sustainability*, 8:12:13–47.

Lyytimäki, Jari (2014) 'Bad nature: newspaper representations of ecosystem disservices,' *Urban forestry and urban greening*, 13:3:418–24. https://doi.org/10.1016/j.ufug.2014.04.005Lyytimäki.

Lyytimäki, Jari and Maija Sipilä (2009) 'Hopping on one leg - the challenge of ecosystem disservices for urban green management,' *Urban forestry and urban greening*, 8:4:309–15. https://doi.org/10.1016/j.ufug.2009.09.003.

Mangiameli, Gaetano (2017) 'Una salutare ambiguità: orti urbani, antropologia, trasformazioni,' *Archivio di etnografia*, 12:1/2:113–37.

Marotta, Gennaro, Cristina Perale and Meme Pandin (2017) *Gli orti sociali a Venezia. Percorso per la gestione unificata e la riprogettazione 2017–18*. Venice: Comune di Venezia. Direzione Coesione Sociale, Servizi alla Persona e Benessere di Comunità. Settore Servizi per la Coesione Sociale.

Millennium Ecoystem Assessment – MEA (2005) *A report of the millennium ecosystem assessment. Ecosystems and human well-being*. Washington, DC: Island Press.

REPAiR (2017) *D3.1 Introduction to methodology for integrated spatial, material flow and social analyses.* http://h2020repair.eu/wp-content/uploads/2018/03/Deliverable_3.1_Introduction_to_methodology.pdf (Accessed August 10, 2022).

Sanyé-Mengual, Ester, Kathrin Specht, Thomas Krikser, Giuseppina Pennisi, Caterina Vanni, Francesco Orsini and Giorgio Prosdocimi Gianquinto (2018) 'Social acceptance and perceived ecosystem services of urban agriculture in Southern Europe: the case of Bologna, Italy,' *PloS One*, 13:9: 1–21.

Schneider, Jane and Peter Schneider (2003) *Reversible destiny. Mafia, antimafia and the struggle for Palermo*. Berkeley: University of California Press.

Silverman, Sydel (1963) *The three bells of civilization*. New York: Colombia University Press.

Smith, Gregory (2012) *La comunità e lo Stato. Antropologia e storia nella Marsica del Novecento*. Luco dei Marsi: Aleph.

Terra Madre (2006) *300 presìdi e 1200 comunità del cibo nel mondo.* Bra: Slow Food Editori.

United Nations (2019) *World populations prospects 2019. Highlights.* https://population.un.org/wpp/publications/files/wpp2019_highlights.pdf (Accessed August 7, 2022).

Vanolo, Alberto (2015) 'The Fordist city and the creative city: evolution and resilience in Turin, Italy,' *City, culture and society*, 6:3:69–74.

4 The landscapes of antimafia farms

Mafia-type criminality has plagued the Italian peninsula starting well before the unification of the nation (Blok, 1975). Its close ties with politics are well established, and indeed collusion between politics and crime groups is a defining feature of mafia (Paoli, 2008). Not by chance the regions with endemic mafia are the poorest in the country, especially the three southern regions of Sicily, Calabria and Campania. Today mafia-type criminality has expanded to other parts of the nation, in a process described as the progressive colonization of the north (dalla Chiesa, 2017). Yet in the south mafia is still present, and impacts many sectors, including agriculture (Corrado, 2011). Giovanni Falcone once commented that the struggle against mafia was like a cart with two wheels, one being the repressive action of the state, the other being the efforts of citizens and associations striving to promote healthy economic relations and robust communities through a culture of legality (Santino, 2015). Social farming belongs to this set of efforts.

A potent instrument for understanding the logic of antimafia farming is the value chain, highlighting process outcome and putting value creation in a perspective that goes beyond financial gain. In Sicily mafia goes by the name of 'Cosa Nostra' meaning 'Our Thing.' They are specialized in capturing control of key economic nodes exploited for the benefit of a restricted criminal elite. Antimafia activists struggle against this distorted broker capitalism, reestablishing control of critical relationships and managing them in the interest of a broader community. Their strategy pursues multiple outcomes and is never based on profit alone. This transformation of social and economic relations is given public attention through communication strategies which herald a better world.

A primary concern for these groups is environmental stewardship, countering criminal activities known as ecomafia, finally addressed in the criminal code by Law 68 of 2015 to punish crimes against the environment. In 2021, the highest court of appeals upheld a conviction of eighteen years handed down to a criminal responsible for helping

DOI: 10.4324/9781003279914-4

The landscapes of antimafia farms 67

to turn the area north of Naples into the Land of Fires (Gaita, 2021). We are speaking of the urban region between the Province of Naples and Caserta, which for a period was the largest illegal dumping site in Europe, covering twenty-five hectares with 2 million cubic meters of waste (Corona and Fortini, 2010). These contaminated sites were extremely lucrative, generating over the years illegal revenues worth €17 billion. They are now being remediated at significant public expense.[1] With almost 4,000 eco-crimes reported in 2019, the region of Campania has the highest number of environmental offences recorded in Italy. Close behind are Calabria, Puglia and Sicily, all areas with a significant history of organized criminality. Yet the list goes on with regions that one would not normally associate with mafia-type criminality, including Tuscany, Lazio and Lombardy. These seven regions claim responsibility for half the environmental crimes committed in 2019. Of all the environmental crimes illegal building is the most persistent, with a 70% increase between 2017 and 2018.[2] Clearly, the defense of the environment is a cogent and timely objective.

Our study of antimafia farms began in 2017 and carried on through early 2022. In pursuing this research, we carried out a series of interviews based on a non-positivistic approach engaging activists with whom we have had sustained contact over many years (Roulston, 2018). The interviews implemented a semi-structured format using questions aiming to elicit reflections on the value chain as an analytical tool (Box 4).

Box 4 Nine research questions concerning antimafia farms

1 What actors belong to your work world, including people, concepts and things?
2 What are the principal stages of your production process understood as a value chain?
3 What actors are involved in each stage?
4 How were these relationships established, and in what way do they resist change? Is the value chain concept useful in discussing your activities?
5 How do you manage innovation?
6 Does climate change impact your activities?
7 What are the intellectual reference points for your enterprise?
8 Is discussion with banks and financial institutions different from other types of interlocutors?
9 What impact have criminal organizations had on your pursuits?

This investigation is part of an ongoing dialogue based on 'the focused selection of the microcosm,' a conversation structured to reflect different points of view (Bobbio and Pomatto, 2007:11). Our study takes into consideration three areas: western Sicily and northern Campania are both noted for their history of organized criminality, while Rome has recently been at the center of attention for what was dubbed Rome Capital of Mafia (Abbate and Lillo, 2015). Our particular interest is in social farming, an activity defined by Law 141 of 2015. The terms of this law are consistent with the orientation law of 2001 (Decreto Legislativo n. 228) which identifies the broad social and economic impact of farming. The idea that social farming can have positive transformational capacity is also found in Law 109 of 1996 which governs the reassignment of confiscated mafia estates and their reuse for institutional and social purposes. The farms in Sicily and Campania explored in these pages operate on such confiscated estates. The farm in Rome operates on underutilized land belonging to the city of Rome. The land is underutilized, claim the activists, precisely owing to speculative forces which they term of a mafia type. Mafia interference in all three cases has wrought damage to farming, the environment, and civil society.[3]

The rise of antimafia cooperatives occurred in a period when quality farm production and attention to specialty products flourished in many parts of the country. Two success stories in Italian agriculture are Campania and Sicily, especially in regard to wines whose commercial achievements are driven by quality and innovation. Campania took the lead in this trend, perfecting local grape varieties once unknown to national and international markets. Investment, innovation and skillful marketing propelled this process of economic growth.[4] Sicily is not far behind, revitalizing a wine industry once limited to low-quality bulk products.

Sicily is the iconic home to mafia, notably in the western valley taking its name from the Belice River. The southern part of this valley is called the Belice Corleonese, a toponym which gives us the Corleone mafia of cinematic fame. The area north of Naples was once the site of celebrated agricultural achievements, seriously damaged in the late twentieth century by mafia activities which include illegal waste management. Campania's mafia-type crime group is the camorra, and its 'Casalesi' branch, rampant in the province north of Naples since the 1990s, is defined as one of the camorra's most stable and powerful clans (Direzione Nazionale Antimafia, 2015).[5]

Rome has its own mafia, which of late infiltrated local institutions and led to convictions under the mafia-specific Law 646 of the Italian criminal code (Pacelli, 2018). In all cases, mafia has deep destructive capacity, and while the state has initiated action against these forces, the role of citizens in strengthening civil society is everywhere held to be fundamental. Social farming and the reutilization of confiscated mafia property plays an important role in this process.

The opportunity to repurpose confiscated antimafia estates started with Law 646 of 1982, promulgated during a series of high-profile homicides committed during the second mafia war. In a dramatic sequence of violence, the main law enforcement officer in Sicily, General Carlo Alberto dalla Chiesa, was killed along with his wife and police escort on September 3, 1982. Such was the public outcry that parliament passed Italy's decisive antimafia law (Law 646) only ten days later. This prompt action helped allay public perception that parliament is complacent when dealing with Italy's oldest justice problem. A critical provision of the law is that the property of mafia criminals can be confiscated and reassigned to associations working for the betterment of the community. Using mafia property against the mafia is a remarkable innovation in the struggle for a culture of legality. Since the passage of Law 646, 26,000 properties have been confiscated. Yet, the management of the reassignment process is far from ideal, considering that only 11,000 of these properties have been reassigned. The rest have been left to sink into dilapidation, depriving the community of important resources. This failure was noted in the early years of the law's application, prompting one of Italy's most important antimafia groups, Libera Association, to circulate a petition asking that a specific law be created to govern the management and reassignment of these assets. Libera is a network of associations, social cooperatives, movements and groups founded by a Catholic street priest in 1995. They collected 1 million signatures to spur on the approval of Law 109 in 1996 on the 'public and social reutilization of confiscated property' (Picciotto, 2015). After years of further failure, a national agency was instituted in 2010 to streamline the reassignment process, thus promoting a highly 'centralized approach with specialized institutions' as a measure to prevent infiltration by organized crime groups (Vettori et al., 2014:24). The subsequent step to reconstruct the regulative framework is given by Law 161 of 2017 revising the code of antimafia legislation and further rationalized the asset recovery process. According to this law, confiscated assets can be reused also for economic purposes, provided their revenues are used for social activities. Finally, Decree 113 of 2018 introduced the possibility of using confiscated assets as social housing.

While these advances have still not fixed all the problems concerning the reassignment of confiscated estates, these legal provisions represent the most advanced set of legislation in Europe in managing confiscated criminal resources and have been emulated all over the continent. As is often the case, difficulties in this sector are not only due to legal measures, but more importantly to the application of these measures. Fortunately, the state's intermittent action is countered by the enthusiastic engagement of so many antimafia associations.

Libera Association continues to play a leading role in monitoring the reassignment of confiscated mafia property and supports extensive research on the topic.[6] These assets are distributed throughout the country. For instance, in Lombardy the confiscated Casa Chiaravalle, a 1,600-square-meter villa, is managed by an association which gives shelter to the homeless (Biancolatte et al., 2017). Rome itself has 500 confiscated mafia properties, almost all residential or commercial real estate. The bureaucratic impediments and poor management that delay reassignment are directly correlated with the 90% failure rate in the enterprises which are assigned these assets (Biancolatte et al., 2017). Examples of failure abound. Frijenno Magnanno was a pizza restaurant in Rome which generated legitimate revenues of half a million euros a year while in mafia hands. After confiscation and reassignment, it went bankrupt within a year. We cannot underestimate the challenges faced by antimafia activists working confiscated property, nor the potent role they play in propelling the nation to a more prosperous future.

Placido Rizzotto

The Placido Rizzotto cooperative is located in the Belice Valley, not far south of Palermo.[7] The valley hosts an ancient agricultural tradition. Historically, it was dominated by Roman latifondi, later transformed into feudal estates, and finally converted into large-scale capitalistic farms in the early nineteenth century. These persisted up to the Agrarian Reform of 1950, after which land was fragmented progressively, in a trend later reversed with the consolidation of smallholding estates. The terrain is naturally arid, although studies show that water shortages are artificially created by corrupt politics and mafia control (Giglioli and Swyngedouw, 2008).

The Belice Valley is known for historical progressive grassroots movements with distinctive socialist resonances dating to the 1890s (Rakopoulos, 2018). Paradoxically, it is also known for the most brutal mafia in Italian history. The early progressive movements were peasant revolts mobilized against what a parliamentary investigation of 1876 called the 'iron circle' bringing together criminality, corrupt politics and major economic interests. The effort to stand up against these forces was coordinated by a movement called the Fasci, referencing the 'bundle' created by united peasants struggling against sinister forces of oppression. This struggle for the right to dignified work is the earliest mass socialist uprising in Europe, defeated by the military force of the state in the years leading up to 1900. This repressive action continued under the fascist government, and the progressive forces only reappeared after the Second World War. They were crushed once again in the massacre of Portella della Ginestra when on May 1, 1947, unidentified gunmen opened fire on unarmed citizens assembled at a socialist rally. The suspicion is that the hidden hand of the state was present here as well. Yet notwithstanding the military scale of repression, progressive forces persist today and are particularly active in social farming.

The cooperative Placido Rizzotto is located in the township of San Giuseppe Jato, a municipality of 8,500 residents. It was once the sovereign territory of Italy's most notorious killer, Giovanni Brusca, who had extensive property in this area when he was captured in 1996. Sovereignty implies autonomy in laws and rules, including the practice of name lending where a mafioso places his property legally in the name of citizens above suspicion to avoid confiscation.

Antimafia farming cooperatives emerged in this area under the umbrella of Libera Association in the early 2000s. Many antimafia groups had their genesis in the late twentieth century, part of a grassroots response to soaring mafia violence. These movements mobilized large segments of civil society starting with Palermo's Spring in 1985 (Schneider and Schneider, 2013). This mobilization started as an urban phenomenon and expanded to rural areas in the late 1990s, especially when property became available to social cooperatives. Libera played a leading role in this transformation, helping manage the reassignment process, and the challenge of turning confiscated estates into viable assets.

Under the terms of the 1996 Law, confiscated property passed to the municipal authorities who then oversaw reassignment to a legally recognized association or cooperative. As today, long-term use rights are given out, while the property continues to be publicly owned. Procedures changed in 2010 with the creation of the National Agency for the Management and Assignment of Property Seized and Confiscated from Organized Criminality. Since then, a magistrate oversees reassignment. In the process involving the cooperative Placido Rizzotto, the city of San Giuseppe Jato created a consortium with another four local townships to coordinate the reassignment of multiple properties. This was the Consortium for Development and Legality.[8] According to claims made by cooperative members today, this strategy not only helped coordinate resources but also reduced the liabilities faced by individual administrators vis-à-vis criminals whose property had been confiscated. Law 109 establishes that cooperatives and associations should be the beneficiaries of reassignment since a major aim of the initiative is to reconstruct civil society. Cooperatives are also said to reduce the risk of mafia racketeering which takes the form of extortion (*pizzo*), imposes the hiring of workers specified by mafia members, the use of specified suppliers, distribution channels and so forth. These burdens leave little leeway for innovation and impose high economic cost. Sharing leadership responsibility among various members is said to help cooperatives avoid this imposition.

The cooperative Placido Rizzotto takes its name from a trade union representative murdered by mafia in 1948. The cooperative was founded in 2001 in a collaborative effort involving the Consortium for Legality and Development, the Prefect of Palermo and Libera Association. Under the prefect's guidance, a public competition was held to select young unemployed workers who could bring back to life 160 hectares of confiscated farmland. Fifteen workers were selected and admitted to a three-month training program. The cooperative received initial financial support from CoopFond, the financial branch of Italy's most important association of cooperatives. These funds were used to repair confiscated agricultural equipment and begin the process of cultivating the land. They also received financial support from the Minister of the Interior's National Operational Program in a project called 'Legality.' It was specifically designed to assist cooperatives working confiscated property.

Today Placido Rizzotto operates an extensive farm, runs a holiday farm (*agriturismo*) called Portella della Ginestra, a horse-riding center entitled to a boy who was strangled by mafia criminals, and a modern winemaking facility. Wines are sold under the commercial name One Hundred Steps, a name taken from a film chronicling the 1978 murder of Peppino Impastato, one of the most noted antimafia activists of the time. One hundred steps expresses the spatial proximity between honest citizens and mafiosi, and the moral journey from criminality to legality. The cooperative is also supported by Slow Food Association.

Some of the initial cooperative members were skeptical about their capacity to yield economic returns and dropped out, then replaced by members with solid farming experience. With this new composition, the cooperative secured a half-million-euro mortgage using personal assets as collateral. These resources were matched by public grants to finance a state-of-the-art winemaking facility, and eighteen hectares of high-quality vineyards. Today the cooperative has a total of twenty-five hectares of vineyards producing 100,000 bottles of wine a year. The winemaking facilities can produce 300,000 bottles a year and use the excess capacity to process grapes grown by other cooperatives in the Libera farming network known as Terra Libera. There are currently nine cooperatives operating in this network in southern Italy, of which four in Sicily. Their wines are sold all over the world using the brand name Centopassi.

Wine is one of the cooperative's most important sources of revenue, although they produce a wide range of agricultural products which are sold through Libera's network of online and offline sales points. The restaurant operated at the agritourism is also important in terms of revenues, leveraging the short value chain (*filiera corta*) which goes from the farm to the fork. Product quality and process outcomes are of fundamental concern, and image is carefully managed, foregrounding their multiple aims. This is especially seen in wine production, where each wine is accompanied by a technical chart showing the name of the mafia boss whose land is used to create the product, and all the other features of the production process. This stresses concern for the environment and local heritage. The antimafia image is important for sales, but the cooperative members recognize that quality is fundamental for continued success. Sympathy may win the first sale, but from there on quality counts in securing a market for their products.

An important outcome of their activities is paid employment, a factor which may be given priority over revenue gains. This was the case, for instance, in their decision to utilize semi-automated bottling equipment rather than a fully automated system. The latter requires the employment of only one worker, while the former requires four. This choice augments production costs but creates work in a region where chronic unemployment has always served mafia interests.

New Organized Cooperation (NCO) and Eureka[9]

The area north of Naples is notorious for organized criminality, associated especially with the town of Casal di Principe which gives its name to one of Italy's most noted criminal groups, the Casalesi. Once called the Campania Felix, or the happy countryside, today this territory is known as Land of Fires. The strong individualism which has long characterized the territory degenerated in the 1980s into a fierce attack on any expression of community solidarity (Berruti and Palestino, 2018). This predatory attitude is visible in the heaps of rubbish and ravaged landscape, as well as in the housing style with isolated homes secured within fortress-like compounds. Community effort to counter mafia met with violence which culminated in the 1994 assassination of the charismatic antimafia priest, Don Peppe Diana. The gravity of this event galvanized community resistance while stirring law enforcement agencies into action.

Image starts with a name. This is strikingly the case in the network of social cooperatives which sprang up after Don Peppe's homicide. The network is organized as a legal consortium bearing the name Nuova Cooperazione Organizzata (NCO). The acronym NCO is best known in this area in reference to a ferocious mafia group dating to the 1980s called the Nuova Camorra Organizzata. The new NCO uses confiscated mafia property against the mafia, as well as the name itself. NCO is today an institutional and territorial brand with high symbolic value, showing how courageous citizens can reverse the territorial decline caused by criminality. The cooperatives support their activities through commercial revenue, but commercial gain is not their chief motivation. They focus on human and social value in a productive system expressing respect for the environment and the territory. They also furnish welfare services, supporting fragile citizens who fall through the cracks of the state welfare system. These are citizens with disabilities who find work and shelter at the cooperatives. These experiments in community health fall under the heading of the 'Health Budget,' a system where the modest resources made available by national health care are combined with resources made available by the cooperatives (Righetti, 2018). Some northern Italian regions have explored this innovative welfare approach, but NCO is the first such experience in the south. All the cooperatives of the NCO consortium are of type B, meaning that at least 40% of their members have documented forms of disability or 'fragilities.'

The consortium brings together various cooperatives operating on confiscated estates, or as they put it estates 'liberated from mafia power' (Natale et al., 2013:1043), and involves a set of seven liberated assets organized as a value chain system.[10] The activities of the consortium are coordinated by the social farm Fuori di Zucca, a pun meaning 'out of your pumpkin,' colloquially a way of saying out of your mind while also referencing agricultural activities. Only crazy people take on the mafia the way they have. The cooperative is located in a former mental hospital, itself a potent symbolic vector. The social farm provides products for retail while acting as a didactic farm for children. They also offer summer camps for kids. Their shop sells products generated by the network of organic farms, including fresh produce, baked products, marmalades, pasta and wine. They also operate a restaurant which uses these products. One of their most ingenious projects is to sell a set of products coming from confiscated estates in a box called *Facciamo un pacco alla camorra*. A *pacco* in Italian means a package, but in the local dialect 'to make a package' means to rip off or cheat someone, in this case the local mafia groups. This project communicates forcefully the philosophy of using mafia assets against the mafia and getting the broader community literally to buy into the effort.

Among the other NCO cooperatives, we find Al di là dei sogni ('Beyond dreams'), Un fiore per la vita ('A flower for life'), Agropoli, ('Farm city'), Eureka and others entitled to mafia victims. In our research, we focused on Eureka since it is chiefly a social farm. They are a small operation working seventeen hectares of land with three permanent cooperative members, a handful of seasonal workers, and half a dozen residential guests who are termed fragile workers. While a core of their property is made up of 'liberated' mafia estates, they also work land belonging to individual cooperative members. Five hectares are devoted to viticulture, twelve to fruit production, mostly peach and pear. Initially they also produced wheat but were unable to make this cultivation commercially viable. They are now exploring the possibility of producing certified organic food products for export to northern Europe. Currently their chief products are wine, pears and peaches.

As with all confiscated estates, starting up is the challenge. The agricultural land they were awarded had not been worked for years, since for mafia owners land is a front used to embezzle European Union funds rather than a farm used for agriculture (Biancolatte et al., 2017). In these cases, one speaks of paper food products, like paper pears or paper oranges. For this reason, the land they were assigned had not been worked for years and was completely overgrown. Making these fields suitable for farming required a major investment of time and resources. Fortunately, the cooperative received public funding from the Minister of the Interior's National Operational Program, used to develop the land as well as their winemaking facilities. They also invested personal resources.

Eureka value chain

Eureka's small operations are organized as a value chain where no step is independent of others. The first step was to bring the farmland back to life. The second was to obtain certification from the local health authorities showing their land was suitable from a sanitary standpoint to produce food for human consumption. This is no small consideration in the Land of Fires. Indeed, all their operations are under close scrutiny by the local health authorities. Some of their property already had fruit trees and some already had vines. They cultivated these plants commercially from the beginning, and with time developed the business model they pursue today. Winemaking has the greatest added value and is their core business. Their winemaking facility is of fundamental importance to their operations and includes bottling equipment for still and sparkling wines. They use two techniques to produce sparkling wines, bulk fermentation and bottle fermentation. Bottle fermentation is more expensive owing in part to labor costs, like rotating the bottles in special racks before recorking. This simple but essential operation is performed by fragile workers who are resident in the winemaking facility. Bottle fermented sparkling wines command a high market price.

They have five hectares of vineyards, and do not anticipate expanding their grape production, since the current volume of grapes saturates their production capacity. This limited-growth model is geared to the needs of their community. While current revenues are sufficient to cover costs, and allow for gradual investments, a major concern is cash flow. They sell out their entire wine production every spring yet must sustain costs all year long. To satisfy cash flow requirements while focusing on their core business, they adopt two distinct business strategies. In the case of wine, where value creation is substantial, they control the entire process, from the vine to the bottle. In the case of fruit, there are limited opportunities to enhance earnings through processing and returns through wholesale are modest. They thus sell their fruit at a low price to a specialized wholesaler while it is still on the tree.

In winter months, cash is tight, and it is then that the cooperative sells their entire pear and peach harvest to specialized intermediaries who manage with their own equipment and labor the final stages of the production process. These intermediaries interact directly with supermarket chains which contract bulk sales through an electronic bidding system. This operation requires specialized commercial capacities which the cooperative members lack, part of a fiercely competitive wholesale market. The bottleneck for farmers is not production but sales. Historically in the Campania Region, the wholesale market was dominated by mafia-type intermediaries, today displaced by impersonal electronic bidding. Both today as in the past this sales structure is classed as a form of vertical integration, where product standards are imposed by the retailer and leave little opportunity for innovation in the production process. Wines sold through the modern retail system face the constraints of a vertically integrated market. But the wine market is more fragmented than that of other agricultural commodities, leaving more opportunities for innovation. This is the reason why NCO have invested particularly in this sector. Something the cooperative members have in abundant supply is innovative creativity. None of the original Eureka members were farmers, but all had a healthy respect for farming. Two of the current members are instead professional farmers, who share a business model balancing between commodity and non-commodity outcomes. They are especially interested in outcomes which promote social integration, preserve the landscape, and maintain local heritage. These aims can only be attained through innovative production processes which guarantee product quality as well as attainment of the desired non-commodity outcomes. Their inclusionary model involves the employment of fragile workers, while also helping disseminate in the local territory the practical benefits of working in a network. Indeed, one of their innovative organizational assets is precisely the capacity to network in a community with strong individualism. This is the core of NCO philosophy and can be extended to many areas. They use this model in their production activities by hiring experts in sparkling wines, paid with their own revenues, and then share out this expertise with other sparkling wine producers in the area. This strategy is foreign to a territory obsessed with individual advantage. Yet the cooperative believes that helping their community of winemakers achieve higher standards of excellence will benefit all producers, including the cooperative.

This area is known for an unusual grape variety, Asprinio, whose high acidity makes an ideal sparkling wine. This grape had long been neglected by commercial winemakers because unknown to the wider market. To revive the grape today is a way to promote biodiversity. Historically, these vines were trained on poplar trees, giving the landscape its characteristic appearance. Working vines trained on these tall trees is not easy, and all vine operations are carried out by specialized workers known as spidermen who use narrow ladders to climb to the lofty summit of their vineyards. Few specialized workers are left, and labor costs are high. But this is the technique used by the cooperative because it satisfies another important process outcome, landscape remediation.

Owing to a fortuitous encounter in 2017 with Camilla Bowles, the Duchess of Cornwall, they now enjoy enormous success in the British market. Since then, they have also begun to explore the American market, and in 2019 started marketing to pizza restaurants on the eastern seaboard of the USA a beverage called Cannaspritz. This is a light unclarified sparkling wine sold at low cost so it can compete for the beer market. Sales were going well until the COVID-19 pandemic broke out, and they had to interrupt their marketing campaign. In 2022, they were gearing up the project once more. These are all examples of how product innovation, marketing acumen and reliance on networks can allow them to reach even remote markets.

Coraggio

Squatting property has an important history in many parts of Italy. This is especially seen in the social centers which sprang into being starting in the 1980s. Most squats identify with progressive politics, including the positions advocated by Direct Action, which as the name suggests proposes the appropriation of essential resources that are not available through other means. A characteristic form of direct action is squatting empty buildings which are used for residential purposes or for cultural activities.

There has been significant debate at many levels concerning how to respond to this practice. The most enlightened debate focuses on the concept of urban commons, seeing underutilized or abandoned real estate as having potential to serve the needs of a wider community (Putini, 2019). This debate has had important institutional ramifications, and some cities have adopted policies to regulate urban commons. The first cities to embrace this initiative were Bologna and Naples. Bologna adopted in 2014 Italy's first comprehensive plan to manage urban commons. In all Italy in 2016, there were 422 properties regulated as urban commons, of which 295 located in Bologna (Putini, 2019:90). While lacking a comprehensive framework, Naples was also among the innovators in this policy area and began to regulate urban commons with an ordinance of 2011. The Neapolitan approach lacks the level of regulation found in Bologna and allows more opportunities for innovative practices.

Rome has still to take action in regulating the use of urban commons. Ignazio Marino began the conversation when he was mayor in 2015, but as of 2022 there had been no progress in adopting a regulation. Standing by is of course a policy choice, considering that Rome has an estimated 400 squatted buildings, and they are rarely evacuated. Some 300 families faced a dramatic episode in 2019 when the local energy company cut off their electricity supply in a building squatted in central Rome, leaving some 450 people in the dark. They had accumulated €300,000 of unpaid bills. With the help of the church the electricity was turned back on.[11] This policy of standing by is a reasonably effective welfare system that entails almost no cost to the city and satisfies the urgent housing needs of many families. There are also dozens of cultural squats which serve the cultural needs of vast urban areas with no public provision of these services. These squats are the Centri Sociali Occupati e Autogestiti (CSOA; Mudu, 2014). Farming squats are instead rare, the most famous being the New Agricultural Cooperative which squatted farmland on the city's western fringe in 1977. They aimed to arrest the city's uncontrolled sprawl and protect its green areas by squatting land for use in social agriculture. After twenty years of squatting, in 1996, they were finally given legal title to the land regulated by an ad hoc agreement with the City of Rome.[12]

Coraggio represents an evolution of these trends, treating unutilized farmland as urban commons and striving to gain legal access without squatting. Coraggio is an acronym meaning courage, derived from Cooperativa Romana Agricoltura Giovani (Young Roman farmers' Cooperative). It was founded in 2015. The activists behind Coraggio come from the underprivileged urban periphery, and experience in the Roman social centers (CSOA). They are sophisticated political activists who have striven to develop a new strategy in handling urban commons. Rome has an abundance of active farmland, but also growing extensions of lost space, created, according to claims advanced by Coraggio, by the speculative pressure of developers. Developers have powerful capacity to influence local politics and shun social farming precisely because it deprives them of a potential investment opportunity. The outcome is that prices for agricultural land in Rome are seven times higher than the EU average. A hectare of arable in Rome sells for €50,000 against an EU average of €7,000, and an Italian average of €17,000. With these prices farming is not a viable investment, allowing developers to create a case for turning vacant land into a lucrative real estate investment. A way to defend these green spaces against encroachment while pursuing environmentally sound outcomes is to treat abandoned green spaces as urban commons. Under this light, Coraggio considers their actions to be those of antimafia activists.[13]

Having been active in Rome's far left movements, the cooperative members had interacted with public institutions over many years, typically from a position of antagonism. They are part of what is referred to as the Roman Movement, historically hostile to public institutions and favorable to direct action. Yet, even within the Roman Movement there are distinctions, with some groups pursuing revolutionary aims, and others seeking accommodation – on their terms – with the local political and economic system. We see this in the squat called the

Little Red House which created a coffeeshop in an abandoned building situated in a poorly maintained park in a fairly central area of the city. They have been highly successful in drawing support from the surrounding neighborhood whose residents regularly use the paid services of the coffeeshop. The activists carried out basic restoration of the building using minimal resources and cleaned up the adjacent green area which now has an extensive seating area. The neighborhood has almost no maintained green areas accessible to the public, and certainly no areas where one can enjoy a cup of coffee in the company of neighbors. By effectively treating the Little Red House as urban commons, the neighborhood has acquired a utilizable green area and a much-frequented public establishment which organizes diverse events, like book presentations, as well as serving excellent coffee, pastries and light lunch products. And all at no cost to the public authorities.

As is often the case in Rome, the absence of a regulatory framework is such that the informal facility of the Little Red House has been tolerated for years by the local authorities. Indeed, they have recently been given informal stewardship over a large adjacent public garden which until then had been abandoned. A gradual process seems to be under way in formalizing the relationship between the squat and the municipal authorities. Historically, given their antagonistic stance, Roman squats almost never had any formal charter establishing legal identity. The activists who run the Little Red House instead operate as a legally empowered 'association promoting social aims,' made possible under Law 383 of 2000. This type of association has fewer constraints than cooperatives or volunteer associations and may be the first step in establishing a legal relationship between the Little Red House and the City of Rome. While squats generally operate under the real threat of eviction, there are many cases where they are transformed into legal occupation, thus coming to represent *ipso facto* an urban commons. Currently the Little Red House operates with no interference, and time will tell if they will eventually be given more legal substance in managing property belonging to the City of Rome.

Like the Little Red House, Coraggio see their actions as a remedy against numerous failings in the current political and economic system. This includes the abandonment of public spaces, the deficit of public services, the high rates of youth unemployment. The support of a broad public is fundamental for the success of these activities, not least because the sale of product and services generates the revenues needed to maintain these activities, both in terms of labor and overhead costs.

Coraggio is unique in this variegated panorama, poised between legality and illegality, between acceptance and rejection of the market logic. They are unique in that they build their activities around urban farming, and the fact that from the start they have had a legal title to the property they occupy. Since 2015, the cooperative has paid rent on a fifteen-year renewable leasehold to twenty-two hectares of farmland owned by the City of Rome. The rent is reasonable, about €6,000 per year. Since 2017, they have received a roughly equivalent EU subsidy for young farmers, a subsidy which is only possible thanks to their legal title to the farmland. But beyond this support, they must finance all improvements to their operations using resources generated by the land itself, since they lack personal assets and resources. In theory, they could seek a loan to improve the property, but this would lock them into the logic of a commercial investment, an orientation they have chosen to eschew. As late as 2022 all the cooperative members had some sort of part-time work outside of the cooperative so they can invest all their farm earnings in improvements. The aspiration is with time to make the cooperative their chief source of personal revenue. The early investments included self-built furniture that could be used to host community events, like wooden tables and benches, rudimentary cooking facilities including an oven suitable for pizza and bread, as well as a small wooden shed used as a sales point for their products. Part of the assignment is an abandoned farmhouse which is being restructured at public expense and should be ready by late 2022.[14] The farmhouse dates to the early twentieth century, constituting the core of a farm known as Borghetto San Carlo. The building will serve as a meeting space, a school for environmental education and a shop for their products.

Assignment of the abandoned farmland came after a long struggle involving the symbolic occupation of the land, endless requests for meetings with city authorities, petitions circulated among residents. The presence of a progressive mayor and regional president in 2015 was fundamental in allowing this activity to culminate in legal access to the land. This assignment opened another phase in the project, namely the investment of labor and funds into the revival of long abandoned farmland. The problem of remedying derelict farmland has been explored in many parts of Europe (Nail, 2008) and the United States (Gelt, 1993). In this case, abandon had caused the soil to become hard and compact, overrun by infesting plants. The activists engaged local farmers in the regeneration effort using heavy mechanical farming equipment to break through the thicket, to plow and till the soil for the first time in decades. The regeneration of derelict farmland averts the risk of desertification and represents an important form of landscape remediation. The land assigned to the Cooperative is situated on a promontory called Borghetto San Carlo, which commands a vista of a well-preserved portion of the campagna romana, rolling hills once given over to wheat, pasture, olive trees and the occasional vegetable garden. This hilltop farmhouse and the surrounding farmland is engulfed by a significant density of residential developments taking advantage of a heavily trafficked artery which leads out to the north. Notwithstanding real estate development pressure, there is still a balance between built areas and protected green including commercial farms. The Borghetto helps integrate the fragmented green corridor which runs along the northern edge of city, with a unique role in providing public access to environmental services as well as the products of the land itself. The cooperative helps preserve a balance between nature and residential communities whose value is defined precisely by these protected green areas. Again, we are in the campagna romana with its iconic Roman pines, including two rows of these trees leading up to the restructured farmhouse. Their poor state of repair is such that a significant investment will be required before the road can be used securely. But they help define a historic landscape which the cooperative wishes to preserve and promote through various activities, including courses on environmental education offered to residents in the school they have entitled to Emilio Sereni.

Crop choice is limited by the absence of water sources for irrigation, and they mostly cultivate drought resistant plants. The exception is a vegetable garden of half a hectare whose produce is sold to the local community at an improvised stand located along the heavily trafficked Via Cassia. In terms of extension, their most cultivated products are found on two and a half hectares given over to hard wheat and sorghum, both of which have limited water requirements. They have an old tractor, but for most mechanical farm services they rely on the paid assistance of neighboring farmers. They use their cereals to make bread, pizza and pasta, sold through their shop. These products are also used to prepare meals organized for community events. In addition to cereals and vegetable crops, they started to plant olive and fruit trees in 2018 thanks to a donation by a local resident. The olive trees are planted on the slopes of their hilltop, yielding olives while preserving the soil against erosion. The wide variety of fruit trees uses diversity to protect the fruit from crop specific pest infestation as an alternative to chemical treatments. All cultivation is organic.

Their revenue comes from local produce sales, and community services including school visits and various types of community gatherings. Before the COVID-19 pandemic, the cooperative organized events with food and live music which brought in as many as 600 visitors from all over northern Rome. According to their estimate, the cooperative is supported by a community of some 300 residents who buy their products and attend their events regularly. Social media maintains this connection. Their business model privileges environmental concern over economic gain, an orientation which caused several members of the cooperative to drop out at an early stage, declining from seventeen members to ten. Where possible they optimize the short value chain to enhance revenue, considering that organic crops, especially cereals, never generate much revenue. Their use of the short value chain relies on networks of like-minded farmers and food processors in other parts of central Italy. Consider the two hectares of arable given over to hard wheat, which costs €500 per hectare to produce while yielding only €800 on the agricultural commodity market. Clearly, wheat production is only advantageous if used for their own processed food products like pasta and bread. But even their short value chain favors healthy practices over profits, driving down earnings. They have their wheat stone milled by a private miller to the north of the city, then transformed into pasta for direct sale to their local public. The cost of producing half a kilo of pasta using this technique is €1.90. Half a kilo of pasta at a supermarket sells for €0.70, while theirs retails at their shop for €3.50. This is a premium price, yet local citizens buy this pasta because they support the cooperative and value the idea of consuming healthy locally produced products. These sales support the landscape, the historical identity of the area, biodiversity, provide work for young people, and maintain a green space which is open to the whole community. This wheat is not a commodity: it is a product of local history, an expression of a living community which supports the multiple outcomes ensured by social farming.

Value chain considerations

The three cases examined here represent examples of social farming used to strengthen civil society and protect the local environment. All these farms rely on well-established networks and strong community support to pursue their aims. The Sicilian case is the most professionalized of the three, benefiting from a well-established network of private and public entities. Their connection with Libera Association in particular allows them to leverage important domestic and international markets. The case of Campania is the most complex in terms of its social mission, pursuing an innovative welfare model harmonized with their agricultural practices. The cooperative in Rome is supported by the community as a bulwark against long-pursued public policies which have harmful social and environmental consequences.

Agriculture links together the diverse dimensions of the material and immaterial world, enhanced by these cooperatives as goals pursued through the logic of the value chain. Community cohesion is a primary goal, but so is respect for the environment, landscape stewardship, biodiversity, cultural distinctiveness and all the positive multifunctional farming outcomes. By analyzing the impact of single moments in the value chain, activists optimize desired outcomes and diminish negative ones. One outcome is financial viability, necessary since these cooperatives receive limited public support. Revenue is greatest with agricultural products processed by the cooperative. Wines are the most lucrative agricultural product, but the wine market is competitive and vineyards are contracting all over Italy. Here we have examples of new entries to the market which enjoy success thanks to quality and innovative choices. Market success for these products is only possible when consumers take the time to explore and understand the multiplicity of farming outcomes. We will see in the next chapter that quality and process outcomes are gaining ever more attention in the Italian food world.

Notes

1 https://www.invitalia.it/cosa-facciamo/sviluppiamo-i-territori/centrale-di-committenza/notizie/terra-dei-fuochi-gara-da-2-7-milioni-per-il-primo-step-della-bonifica / (Accessed December 10, 2021).
2 https://ilgiornaledellambiente.it/la-legge-realacci-efficace-per-combattere-le-ecomafie/ (Accessed December 10, 2021).
3 The idea that a social economy can transform a criminal economy has received recent scholarly attention in Caggiano and De Rosa (2015).
4 There is an ample literature on innovation within Campania wines. An interesting study of business challenges for Campania wine producers is found in Rossi et al. (2012). Buondonno et al. (2006) provide a revealing study of the specific linkage between local grapes and the local territory in the area north of Naples discussed in this paper.
5 We use the term 'mafia' as a blanket expression to cover organized criminality, but note that the Campania mafia is usually referred to as camorra.
6 https://www.confiscatibene.it/chi-siamo (Accessed December 11, 2021).
7 Rakopoulos has published a series of articles (especially 2013, 2014, 2015) on farming and cooperatives in the Belice valley from which this section draws. The section also draws from the web site of *Centopassi*.

The landscapes of antimafia farms 85

https://centopassisicilia.it/it/ (Accessed July 17, 2018). Gregory Smith had various occasions to interview the head of the winemaking facilities, Stefano Palmeri, between fall 2017 through spring 2019.
8 http://www.sviluppolegalita.it/cooperativa-placido-rizzotto-libera-terra/ (Accessed July 17, 2018).
9 Both authors have had occasion to conduct research with the cooperative over the last several years.
10 The cooperatives involved in the consortium include: Eureka, Al di là dei sogni, Un fiore per la vita and Agropoli. The authors carried out a series of interviews of consortium and cooperative members starting in the summer of 2018 (Giuliano Ciano, Giuseppe Pagano, Pasquale Corvino, Vincenzo Letizia).
11 https://www.ilmessaggero.it/roma/news/palazzo_occupato_spin_time_labs_papa_corrente-4486612.html (Accessed December 18, 2021).
12 http://www.agricolturanuova.it/storia-2/ (Accessed December 18, 2021).
13 Gregory Smith had various opportunities to interview cooperative members, especially the head of the cooperative, Giacomo Lepri, starting in summer 2018.
14 The restoration is being undertaken by a developer as a way of compensating for the investment tax which is required for a real estate project authorized in another part of the city. This is called an *opera a scomputo*.

Bibliography

Abbate, Lirio and Marco Lillo (2015) *I re di Roma. Destra e sinistra agli ordini di Mafia Capitale*. Milan: Chiare Lettere.
Berruti, Gilda and Maria Federica Palestino (2018) 'Looking from Italy at the fertile boundary between formality and informality,' *CRIOS. Critica degli ordinamenti spaziali*, 15:27–40.
Biancolatte, Tecla, Gianluca De Martino, Daniela Larocca and Jacopo Salvadori (2017) *Riprendiamoli - Inchiesta sui beni confiscati alla mafia*, Kindle ed. Rome: Gedi News Network.
Blok, Anton (1975) *The Mafia of a Sicilian village, 1860–1960: a study of violent peasant entrepreneurs*. New York: Harper Collins Publishers.
Bobbio, Luigi and Gianfranco Pomatto (2007) 'Il coinvolgimento dei cittadini nelle scelte pubbliche,' *Meridiana*, 58:45–67.
Buondonno, Andrea, Abner Colella, Elio Coppola, Bruno de Gennaro, Alessio Langella and Carmine Colella (2006) 'A proposal for an eno-geopedological map of Campania region (Italy) based on vineyard/tuff formation links, a study case on "asprinio" terroir,' in *Proc. Conv. Int. Paesaggi, terroirs e i paesaggi del vino*. Perugia, pp. 6–9.
Caggiano, Monica and Salvatore Paolo De Rosa (2015) 'Social economy as antidote to criminal economy. How social cooperation is reclaiming commons in the context of Campania's environmental conflicts,' *PACO-The open journal of sociopolitical studies*, 8:2:530–54.
Corona, Gabriella and Daniele Fortini (2010) *Rifiuti. Una questione non risolta*, Rome: XL Edizioni.

Corrado, Alessandra (2011) 'Clandestini in the orange towns: migrations and racisms in Calabria's agriculture,' *Race/ethnicity: multidisciplinary global contexts*, 4:2:191–201.

dalla Chiesa, Nando (2017) *Passaggio a Nord: la colonizzazione mafiosa*. Turin: Edizioni Gruppo Abele.

Direzione Nazionale Antimafia (2015) *Relazione annuale sulle attività svolte dal Procuratore nazionale antimafia e dalla Direzione nazionale antimafia nonché sulle dinamiche e strategie della criminalità organizzata di tipo Mafioso*. http://www.camera.it/temiap/2015/03/04/OCD177-1033.pdf (Accessed July 15, 2018).

Gaita, Luisiana (2021) 'La Cassazione conferma la condanna a 18 anni per Cipriano Chianese, "l'ideatore delle ecomafie" per conto del clan dei Casalesi,' Rome: *Il fatto quotidiano*, January 20.

Gelt, Joe (1993) 'Abandoned farmland often is troubled land in need of restoration,' University of Arizona Water Resources Research Center (WRRC). https://wrrc.arizona.edu/publications/arroyo-newsletter/abandoned-farmland-often-troubled-land-need-restoration (Accessed July 14, 2022).

Giglioli, Ilaria and Eric Swyngedouw (2008) 'Let's drink to the great thirst! Water and the politics of fractured techno-natures in Sicily,' *International journal of urban and regional research*, 32:2:392–414.

Mudu, Pierpaolo (2014) 'Where is culture in Rome? Self-managed social centers and the right to urban space,' in Isabella Clough Marinaro and Bjørn Thomassen (eds.) *Global Rome: changing faces of the Eternal City*. Bloomington: Indiana University Press, pp. 62–78.

Nail, Sylvie (2008) *Forest policies and social change in England*. New York: Springer.

Natale, Alfredo, Caterina Arcidiacono and Salvatore Di Martino (2013) 'From "Gomorrah Domain" to "Don Peppe Diana Lands." A southern Italian experience of work-based liberation. Community networking and well-being,' *Universitas Psychologica*, 12:4:1039–49. https://doi.org/10.11144/Javeriana.UPSY12-4.fgtd.

Pacelli, Valeria (2018) 'Mafia capitale era Mafia,' *Il fatto quotidiano*, September 12.

Paoli, Letizia (2008) *Mafia brotherhoods: organized crime, Italian style*. Oxford: Oxford University Press.

Picciotto, Loredana (2015) 'The social innovation of social cooperatives operating on confiscated mafia properties: support factors and critical issues,' in *Managing intellectual capital and innovation for sustainable and inclusive society. Proceedings of the Make Learn and TIIM Joint International Conference*. Bankok: ToKnowPress, pp. 543–58.

Putini, Antonio (2019) *Beni comuni urbani. Soggetti, pratiche e retoriche della città condivisa*. Milan: Franco Angeli.

Rakopoulos, Theodoros (2013) 'Food activism and antimafia cooperatives in contemporary Sicily,' in Carole Counihan and Valeria Siniscalchi (eds.) *Food activism: agency, democracy and economy*. London: Bloomsbury Academic, pp. 123–42.

Rakopoulos, Theodoros (2014) 'Cooperative modulations: the antimafia movement and struggles over land and cooperativism in eight Sicilian municipalities,' *Journal of modern Italian studies,* 19:1:15–33.

Rakopoulos, Theodoros (2015) 'Which community for cooperatives? Peasant mobilizations, the Mafia, and the problems of community participation in Sicilian co-ops,' *Focaal-Journal of global and historical anthropology,* 71:57–70.

Rakopoulos, Theordoros (2018) *From clans to co-ops: confiscated mafia land in Sicily,* Oxford: Berghahn Books.

Righetti, Angelo (2018) 'I "Budget di salute", progetti terapeutici contro la finanziarizzazione del welfare,' in Antonio Esposito (ed.) *L'atlante delle esperienze di riutilizzo e mancato riutilizzo dei terreni confiscati e delle realtà di agricoltura sociale in Campania.* Soveria Mannelli: Rubbettino, pp. 365–80.

Rossi, Matteo, Demetris Vrontis and Alkis Thrassou (2012) 'Wine business in a changing competitive environment–strategic and financial choices of Campania wine firms,' *International journal of business and globalization,* 8:1:112–30.

Roulston, Kathryn J. (2018) 'Qualitative interviewing and epistemics,' *Qualitative research,* 18:3:322–41.

Santino, Umberto (2015) *Mafia and antimafia: a brief history.* London: IB Tauris.

Schneider, Jane and Peter Schneider (2013) *Reversible destiny: Mafia, antimafia, and the struggle for Palermo.* Oakland: University of California Press.

Vettori, Barbara, Tudor Kolarov and Atanas Rusev (2014) *Disposal of confiscated assets in the EU member states. Laws and practices.* Sofia: Center for the Study of Democracy.

5 Alternative value chains

In Italy, the distinction between alternative food networks (AFN) and conventional food procurement requires careful definition (Sonnino and Marsden, 2006). AFNs are a response to impersonal globalized agri-food markets, to which even large-scale Italian supermarkets advocate opposition. The short food chain (*filiera corta*) is one element in defining AFNs, but by itself is not enough. AFNs strive to re-localize and re-spatialize food relations, embedding them in personalized social relationships. They are territorial systems deploying spatial strategies, resources and social relations which are in contrast with the mainstream food system (Dansero and Putilli, 2014). AFNs build close spatial connections between production, distribution and consumption. Resources include immaterial assets like knowledge and skills, as well as the material factors of production. Both types of resources are typically under local control or embedded in networks of like-minded AFN activists. The social relations are personal, with a distinctive element of participation and democratic decision making.

Coproduction is a critical concept, where consumers seek informed knowledge of food production factors in a dialogue between consumers and producers (Petrini, 2005). The success of sustainable AFNs is determined by consumers who go beyond surface appearances. This is the case of most Italian food shoppers, who have never been completely captured by supermarket culture, and seek informed knowledge about their food using consolidated personal networks. This makes Italy's food system unusual from the standpoint of western European transformations, with the persistence of fragmented food networks and the delayed emergence of large-scale retailers. Box 5 provides selective numerical information on the Italian food procurement system.

DOI: 10.4324/9781003279914-5

Box 5 Selected numbers concerning the Italian food system

In 2021 the total food system in Italy was worth €575 billion. Exports were equivalent to €52 billion. In 2018 there were 175,655 food sales points employing a total of 734,273 workers, with an average of 4.18 workers per point.

Supermarkets

Supermarkets constitute (2021) an increasing percentage of food sales in Italy, equivalent to some 60% of total food retail. There is a relatively low concentration of large supermarket groups in Italy where the top five account for 58% of all supermarket sales as compared to 80% in France.

Solidarity purchasing groups (GAS)

Some 12% of Italians make regular use of this system of food provision.

Internet sales

It constitutes 3% of total food sales in Italy as compared to 13% in the USA.

Sources:
ISTAT (2021) *Annuario statistico italiano 2021*.
Rome: ISTAT
Coldiretti Economia. https://www.coldiretti.it/economia/covid-record-filiera-cibo-a-575-mld-7-nel 2021#:~:text= Nel%202021%20il%20cibo%20diventa, le%20difficolt% C3%A0%20legate%20alla%20pandemia
(Accessed August 14, 2022).

Since the 1950s, the European trend has been characterized by the supermarket revolution and the vertical coordination of the agri-food market (McCorriston, 2006). Vertical coordination binds the producer to a transforming industry, wholesaler or retailer, increasing revenues for contracted farmers while determining a loss of competitiveness for smallholding farmers and undercapitalized farms. Vertical coordination is definitionally impersonal, generally insensitive to territorial concerns as well as to ethical ones. Persisting horizontal integration is what gives Italy its special character, a fragmented market allowing greater interdependence among multiple actors typically bound together by ongoing personal ties.

Slower Italian transformations evident in these complex food networks are seen in the coexistence of various food provision systems, including neighborhood produce markets, farmers markets, small shops and a host of AFNs. These systems are embedded in communities all over the country, an expression of the ubiquitous concern for food quality and distinctiveness (FIVA, 2017). In the postwar period, a persisting feature of the Italian agri-food system has been neighborhood markets (*mercati rionali*) retailing food and nonfood products and services. These markets are often the physical core of a neighborhood, selling fresh produce, fresh baked products, meat, fish, wine and so forth. Non-food items include cleaning products, clothing and services like shoe repair. They often have coffeeshops, and other establishments serving prepared foods, as well as commercial services like real estate agencies, and occasional non-commercial services like time banks. Neighborhood markets draw attention to the social dimension of the food system and countervail the mass-produced market by bolstering awareness of food production factors. Personal relationships between buyers and sellers are maintained over years if not generations, building trust relationship which disseminate precious information about food quality (Black, 2014). Sharing information at this intimate range strengthens local knowledge while sustaining local distinctiveness. All types of information are shared, including recipes involving seasonal products, helping to maintain the fine-grained territorial distinction so characteristic of the country.

Today, neighborhood markets face major competition from supermarkets, although a study commissioned by the Italian Federation of Street Vendors shows that the relationship between the two is synergetic rather than antagonistic. While Italian shoppers turn to supermarkets for price and convenience, they still frequent neighborhood markets regularly. According to this study 90% of Italians make use of a neighborhood market at least once a week, underscoring the territorial character of the Italian food system (FIVA, 2017). These markets are supported by Italian municipalities as a fundamental strategy in ensuring food security. Their salience in the daily food system is such that supermarkets emulate their characteristics, striving to build personal ties while expressing concern for regional distinctiveness, respect for the producer, attention to the environment. A major supermarket chain, Conad, uses the slogan 'people behind the things' as an appeal to the logic of trust found in traditional markets. Neighborhood markets today are staffed chiefly by retailers; they are not farmers markets in the strict sense. The few farmer-retailers that persist in neighborhood markets publicize their special status as a guarantee of quality and distinction. Most persisting farmer-retailers today operate stalls in the many farmers markets strictly speaking that have emerged in the past twenty years using the English language name 'farmers market,' or the Italian equivalent *mercato contadino*. These markets generally meet once a week as opposed to the conventional neighborhood markets which meet six days a week. The separation between production and retail is a major change in the food retail system, made necessary owing to declining family size. In the early postwar period, families were large and could usefully diversify by having some members work the farm, some staff retail points. The 1980s marked a shift in family size, with small units which lack the resources for this form of internal diversification.

Up to the modern period, Italy was a land of small- or medium-sized towns, expressing the small-is-better idea championed by the great nineteenth-century writer Carlo Cattaneo (1965). Cattaneo believed that all the amenities of the city should be accessible to citizens in a short walk, an idea preserved today in the fifteen-minute city concept. The modern community ideal draws from the historic concept of the neighborhood unit now applied to contemporary planning, especially in the post-pandemic era (Manzini, 2021).[1] Even in large cities, the pedestrian concept is supported by planning concern for neighborhood-level services, ensured by combining residential and commercial functions integrated with public services like health care and schools. Conceptually Italian cities have long been characterized by a sharp rural–urban divide, yet connections between these two spheres are always strong, allowing farmers to benefit from the discerning insights of consumers linked to wider territorial networks, and consumers to benefit from high-quality distinctive agricultural products (Montanari, 2008). Today Italy is less provincial, and life is lived at a greater distance from the territory. Territories are in constant transformation due to changes in urban fabrics, as well as demographic and economic dynamics. Urban and rural aspects interweave, especially in the urban fringes surrounding metropolitan cities. These are peri-urban areas 'consisting of urbanized areas beyond the dense and consolidated city, crossed by differentiated phenomena of settlement expansion beyond the limits of the countryside, which identifies rural and open space, traditionally coinciding with the limits of the city' (Russo and van Timmeren,

92 Alternative value chains

2022:17). In peri-urban territories, attention to circularity and regeneration are today critical, complementing an ongoing process of urban regionalization (Soja, 2011). An important element in regenerating the contemporary relationship between the rural and urban are AFNs which have arisen in direct response to the loss of local character in some sectors of the modern food system. Yet notwithstanding many transformations, traditional practices and values persist and are in some ways still written into the fabric of the city.

Neighborhood retail

Rome can be taken as an example of how retail is organized at the neighborhood level. Its history of informal growth is such that neighborhoods have often sprung up in a way dictated by the habits of residents rather than any planning logic (Smith, 2021). Rome is known as an archipelago city comprising some 200 neighborhoods having an average population of 15,000 inhabitants. Twenty years ago, each had its own neighborhood market, a space which shaped community identity while satisfying household shopping needs. An example of the local food retail system is Torre Maura, on the city's eastern fringe. It was a rural settlement as late as the 1920s and has today grown into a compact neighborhood home to 9,000 citizens.[2] The major period of growth was in the 1950s, and by 1960 the neighborhood was sufficiently consolidated for a produce market to emerge. Like most markets in Rome, it started as an informal food retail hub with non-permanent stands, and gradually evolved into a neighborhood market with more robust facilities coming under the administrative control of the city's commerce department. An aerial view of this area shows that the neighborhood is centered on the market which runs along a wide central street, a spontaneous creation rather than a purpose-built market. Though lacking the formal design features of a public space, this is the core for the neighborhood's public assembly. Three streets intersect at this node, where the market, comprising some thirty stalls, operates six days a week from 7 am to 2 pm. The market sells fresh produce, cheeses and cold cuts, meat, fish and dry goods. Cut flowers are on display in a central stand, drawing the passerby's gaze. The spaces are now officially designated, hosting mostly permanent street vendors, but with two spaces for rotating vendors with non-permanent stands who sell clothing, shoes, handbags and accessories. Most of the neighborhood is within a fifteen-minute walk of the market.

According to recent research, 40% of the food sold in Rome's neighborhood markets comes from the Italian south, 25% from the Lazio region, 20% from northern Italy. The remainder, about 15%, is imported from a foreign source.[3] At Torre Maura, most vendors are native Italians, although some stalls employ immigrant vendors, especially Bangladeshi who are so prominent in Rome's food retail system (Knights, 1996). The market constitutes the center of the neighborhood's service area, with two banks, three coffeeshops, a computer repair shop, a nail repair shop, two jewelry shops, a shoe store, a cell phone store and a Conad supermarket. The supermarket has a sales area of about 250 square meters. There are also two dry goods stores in the area, one operated by a Chinese family the other by Italians. Chinese dry goods stores are found all over Italy, usually situated in proximity of the neighborhood market, and sell a complete range of household products, clothing and hardware items.

With 183,000 street vendors, Italy is Europe's leader in this type of retail. Almost all vendors are individual firms, with close links to their client base and no form of vertical constraint. An increasing number of these vendors are of foreign descent, now more than half the total.[4] But the immigrant vendors like the Italians make a point of developing strong personal ties with their clients, a fundamental capacity for the success of the firm. Almost all stalls sell products which are identical to those sold at other stalls, discouraging competition on price since earnings are already modest. In this setting, competitive advantage is determined by the trust relationship established between merchant and client. According to one estimate, the earnings of a stall keeper are €1,000 net per month, in a country where the average salary is about €1,500 a month.[5] Independence and interaction with the public are the chief sources of satisfaction for these workers.

We have noted that these are retailers rather than producers, and stall keepers typically buy their produce daily at the municipal wholesale market which begins its operations at 5 am. Most cities in Italy have wholesale food markets to serve the needs of this system, like the CAR in Rome, the CAAN of Naples, or the SOGEMI of Milan. Regional producers sell their products there daily, allowing stall keepers to boast that their products are harvested the day before they are sold on the stands. No special packaging is required to ensure food security.

Naples has fifty neighborhood markets,[6] Milan reports ninety-four markets, or one per 14,000 population.[7] Florence has twenty-nine markets, or one per 13,000 population.[8] The city of Venice reports the presence of seven markets in the historic center and islands, or one per 7,500 population.[9] There is high turnover of stall keepers today, about 10% a year, partly because of European and national legislation

94 Alternative value chains

which makes it relatively easy to obtain a vendor's license. Previously a restricted number of licenses was issued and purchased by individual merchants at significant cost. In those days, before changes in the 1990s, the flourishing activity of neighborhood markets combined with the security afforded by the limited issue of vendors' licenses provided a form of security and an incentive to continue working in this sector over many years. The EU started to encourage more freedom in commercial activities starting in the 1990s, culminating with the Bolkestein Decree of 2006. This decree makes it easier for licenses to change hands, encouraging competition while diminishing the security of license holders. Understandably existing license holders are completely opposed to this decree, and its effects have been subject to continual postponement carrying up to 2021.[10] Historically, the limitation of licenses was such that a license could be sold on an informal market after many years of service generating a lump sum that the stall keeper could expect as a kind of severance indemnity at the end of a working career. With the liberalization of licenses, there is no longer any form of indemnity. On the positive side, however, the liberalized system greatly reduces entry barriers, making it possible for new operators to access the system. The ease with which licenses are available today, among other things, has opened the way for immigrant stall keepers with limited resources.

Retail food procurement trends are shifting, as shoppers become used to sales points with long opening hours and easy access by car, two trends which have encouraged the growth of supermarkets. Neighborhood markets are still an important component of the food retail system but must increase revenues by creating new activities on their premises. In Rome, a well-known initiative is the Mercati di autore, an association of markets which has introduced non-food activities like book stalls, book presentations, lectures and debates. Some markets now sell high-end specialty products catering to the taste of an affluent public. Increasingly markets also sell prepared food and offer seating areas especially at the lunch hour. Markets are an expression of the versatility of the food system, optimizing positive multifunctional outcomes. They represent chiefly the short value chain, assuring high standards in terms of personal health, product quality and the ethical features of food production.

Supermarkets

The first supermarket in Italy was the Long S, founded in Milan in 1957. This was Italy's first step toward transforming food into a commodity, and coincided with Italy's first spot ad on television, in a program called Carosello.[11] The Long S was taken from the word supermarket, and the slogan was 'a long S with short prices.' To compete more effectively with neighborhood markets, the Long S created in 1974 Italy's first supermarket gastronomy sales counter, imitating specialized neighborhood market stalls that sell cold cuts and cheeses prepared to order.[12] Over time supermarkets have catered increasingly to automobile mobility, and mega supermarkets have developed in all parts of the country. Reliance on car mobility militates against neighborhood markets that are almost always located in dense residential areas with limited parking facilities. But supermarkets are also found in the vicinity of traditional neighborhood markets, often taking the place of the unaffiliated local grocer. This arrangement allows the shop owner to benefit from some of the price advantages of a supermarket chain while maintaining a degree of autonomy in choosing products for sale which represent local neighborhood tastes. Fifteen percent of all supermarkets comprise these small-scale local stores with a sales area of less than 250 square meters. Small scale retail services also enjoy tax advantages aiming to support the neighborhood retail system.

Supermarkets present themselves to the Italian public as a continuation of the traditional food retail system built on trust and local roots. Two of Italy's biggest supermarkets are cooperatives, both with operational headquarters in Emilia-Romagna, a region with unique experience in agricultural cooperatives. Both were founded when Italy was transitioning from a chiefly agricultural economy to greater employment in the industrial sector. Conad was founded in 1962 and today has 65,000 employees and revenues of more than €15 billion. Coop was founded in 1967, has 56,000 employees and revenues of €13 billion. Both aim to modernize the trust relationship typical of Italy's traditional retail system and do so by expressing strong commitment to high standards of ethical behavior and environmental sustainability. All their publicity material stresses these elements both in domestic operations as well as activities carried out in other parts of the world. Coop for instance, publishes ongoing notice of social and environmental campaigns carried out in Italy, like food donation to people in need, support for environmentally friendly suppliers, and urban reforestation. They also give notice of activities carried out in other parts of the world, like COVID-19 vaccine campaigns in Africa.[13]

This is an example of corporate social responsibility (CSR) in the agri-food system. The EU has supported CSR since the early millennial years, with specific interest in the food sector. EU directives and national legislation provide various incentives for corporations to

96 Alternative value chains

adopt actions which are consistent with the new consumption paradigm. Success in achieving these goals is seen in the fact that Coop was rated in 2010 Europe's most socially responsible food corporation and one of the best in the world.[14] Local food retailers have always seen themselves as the interface between complex agricultural relations and the final consumer, a guarantor of quality. Supermarket chains like Coop and Conad have carried this idea forward, positioning themselves as more than purveyors of commodity food. Cooperatives can foreground the importance of social relations in their operations in a way which shares some of the characteristics of an economy based on small firms (Boccia and Covino, 2018). The Italian government supports CSR by cofinancing through tax incentives projects which highlight attention to quality, safety, ecological and social impact. All these themes are important to modern food consumers. The study quoted above found that 97% of Italian consumers have little or no trust in corporations, providing a powerful incentive for large firms to highlight their social and environmental commitments. Contemporary supermarket shoppers are more demanding than in the past. They make purchases based on a good price/quality ratio, but also express concern for the social and environmental consequences of their shopping habits.

A dark page in recent supermarket history was written by the practice of the double lowest bid where farmers are invited to present bids for products in two different stages.[15] The supermarkets that practiced this procurement procedure would put out a tender for a given product, and the lowest bid in the first tender became the starting price for the second tender. This double-lowest-bid process ensured that the final price would be as low as possible, often below production cost. The only way to win the bid was by cutting costs, often illegally. Farmers can augment agricultural yield by applying unauthorized quantities of fertilizers and pesticides, and they can save cost by using underpaid labor or undocumented workers. The exploitation of agricultural labor came to public attention in 2010 when African migrant workers in the Italian south took to the streets to protest against this extreme exploitation (Corrado, 2011). Workers were being hired through intermediaries specialized in recruiting under-cost labor. The response to the public debate generated by this uprising was a law passed in 2016 which introduced significant sanctions for the use of labor intermediaries. Farms using this labor were often aligned with criminal groups, who were also expert in negotiating the double lowest bid procedure. Initiatives taken against this bidding process reached a peak after news was released that one supermarket, the Italian discount group

Eurospin, purchased one million bottles of pureed tomato for the price of 31.5 cents per bottle, then retailed at the price of 39 cents. Yet the cost of tomatoes, the bottle and the cap, excluding labor, was 32 cents per bottle. The only way to sell under 32 cents was to underpay labor.[16] The double lowest bid process was finally prohibited by a law passed in Italy in 2021 implementing the 2019 EU directive on unfair trading practices in business-to-business relationships. Subsequently Eurospin published on its website the claim that it did not use the double lowest bid process, specifying its commitment to the quality of its products and its relations with suppliers.

Supermarkets in Italy represent by far the most common channel families use to acquire their food products. Yet broad consumer concern for social and environmental outcomes requires supermarkets to structure their activities accordingly.

Farmers markets and Campagna Amica

Farmers markets, in the strict sense of markets where farmers sell their produce, have always existed in Italy.[17] The advantage to the farmer in selling at their own market is to optimize revenues by shortening the value chain. But the collapse of family size means that this is not an activity which farmers can perform daily. The only way to reconcile production and distribution within a single-family farming unit is to organize markets that meet once or twice a week. This organization is consistent with modern shopping patterns, where most food shoppers have stopped buying fresh food products daily. The ideal is still daily shopping, but this is rarely done owing to the multiplicity of demands on family time (FIVA, 2017).

The success enjoyed by farmers markets can be attributed to their ability to reconstruct the relationship between the city and the countryside which has been increasingly disrupted by the modern food system. Their structure of meeting weekly or monthly fits well with contemporary food shopping patterns. Thirty million Italians, half the entire nation, buy food directly from the producer at least once a month, and 130,000 Italian farmers rely prevalently on direct sales to the public. Freshness, quality, flavor, food security, support for the farming community, the expression of local identity and the opportunity to socialize with other shoppers motivate this practice. Concern for these factors make Italy the world leader in farmers markets.[18]

Farmers unions have played an important role in launching farmers markets. They know the business and appreciate that direct sales can increment farm revenue. The most extensive network of farmers markets is organized by the Coldiretti union under the banner of Campagna Amica, literally the friendly countryside. Campagna Amica supports farmers markets, holiday farms (*agriturismi*), restaurants and community gardens. Today this network includes 7,200 farms, 1,250 markets, 2,200 holiday farms, 550 restaurants, 210 community gardens, thirty street food sales points, selling products produced on 200,000 hectares of affiliated farms.[19] The business is worth €3.5 billion. From the standpoint of revenues and visibility, these are the most important alternative food networks in Italy. They are closely linked to the local territory, express continuity between urban and rural environments, build personal relationships between producers and consumers, and leverage locally available resources in a way that benefits both farmers and ordinary citizens.

Value chain experiments

The idea of food miles was given life in the United Kingdom in the 1990s with studies correlating food miles to health.[20] Global food products travel enormous distances, and to bring food down to a territorial scale, food activists in Italy created the term zero-kilometer (0 km) food. This term was first recorded in Bologna in 2005 and was soon taken up by Carlo Petrini as a banner for Slow Food.[21] A closely allied concept is the short value chain, today a widely used concept applied to such disparate sectors as furniture and fashion.[22] The idea is firmly established in Italian public awareness, with beneficial implications for social and economic equity, and for environmental sustainability.

The spatial dimension of the short value chain is a restricted territory containing all processes of production, transformation, distribution and consumption. Linkages among the various phases are personal, and the resources involved are under local control. An instructive example of an innovative short value chain operation is Italy's most recent alternative food coop, the Coop Camilla. It opened in Bologna in 2018 as an experiment in coproduction aiming to undermine the logic of the mainstream agri-food food system, furnishing local, natural and ethical food. They refer to themselves as a community emporium, referencing the presence of a territorially restricted group which controls all phases of value generation. In this system, consumers are co-producers, participating in food distribution by providing free labor a few hours each week. This gives consumers access to food which is cheaper and healthier than food sold at conventional retailers. They started with 200 coop members and doubled that number within the first year. Their activities are not restricted to local foods,

Alternative value chains 99

and an important success factor is having considered that the Italian regional diet requires products which are not locally available. Even in historic times, Italians were not strictly locavores, and today many progressive writers recognize the need to go beyond a strictly local food system (Clancy and Ruhf, 2010). In the light of this consideration, Coop Camilla created a network of suppliers from different parts of the country as a restricted group bound together by personal ties and a shared concern for ethical, social and environmental sustainability. Fish, for instance, in particular tuna and anchovies, are supplied by a small-scale producer in Sicily. Oranges are also procured from a small organic producer in Sicily. Sheep products come from the Italian Alps, pasta from the Italian south. They even have soaps and household detergents that are produced organically in Italy, as well as cosmetics. It is a personalized network of like-minded producers organized by a community of buyers who staff their own operations. To date, this is the only experiment of its type in Italy, deploying an idea of the territory not based on physical contiguity alone but also shared relationships, knowledge, values and resources.

Coop Camilla represents an innovation drawing from the experience of solidarity purchasing groups (GAS) which emerged in the 1980s, creating personalized relationships between producers and consumers. The elimination of intermediaries narrows the knowledge gap and eliminates a cost with benefits for both parties. These purchasing groups also create a bond of empathy between the farming community and the urban population. GAS started out as a set of informal exchanges where producers created neighborhood groups of buyers, often organized as condominium groups. These informal groups eventually coalesced as participatory networks constituting a social movement based on the solidarity economy. In 2002, a meeting was held in Verona bringing together the experiences of GAS groups from all over the country. The meeting led to the drafting of the Charter for the Italian Network of the Solidary Economy (RES).[23] The key features of the solidary economy are cooperation and reciprocity, environmental sustainability, local grounding, participation, networking and social utility. They represent a direct relationship between producer and consumer. The Charter defines solidary economy districts whose local character brings together producers, consumers, local administration, financial groups and other actors.

By 2022, there were over 550 GAS networks distributed all over the national territory.[24] A survey conducted by the Observatory for Ethical Consumption in 2020 revealed that in twenty years the number of people involved in GAS networks had increased by 200%.[25] The number continues to increase owing to the 'Greta Effect.' According to this study Italians who identify as ethical buyers account for two thirds of the population. As in many cases of Italian innovative institutional arrangements, we find a gradual formalization of alternative networks. Since 2008, the state has recognized GAS networks as nonprofit juridical subjects. Yet, many groups have resisted the effort to provide them with institutional support. This was seen in the Abruzzo in 2016, when the GAS networks rejected an offer of support from the regional government, claiming that this would reduce flexibility and increase bureaucratic burdens. Public sector financial support was superfluous since the system was economically self-sustaining. Furthermore, public funding would entail an element of formal control which would reduce their capacity to engage in spontaneous innovation driven by the orientation of the participants who make up the GAS networks.[26]

Flexibility and democratic decision making are the fundamental characteristics of GAS networks. However, while information and points of view are shared between the producer and the consumer in this system, production risk falls exclusively on the shoulders of the producer. Various experiments have attempted to narrow this gap, sharing risks as well as choices. These experiments fall under the heading of community supported agriculture (CSA), a model developed in Japan in the 1970s following theories elaborated by Rudolf Steiner in the 1920s.[27] Bologna, Siena and Rome are the leaders in this initiative. In Bologna, ARVAIA has operated since 2007 on a 47-hectare farm on the western side of the city. The farm distributes its products through eight pick-up points scattered throughout the city where their produce is delivered to coop members. They also use these as sales points for the general public. Prices to coop members are based on bulk products, established by estimating the cost of production for a given year. In 2021, the cost was €750 for the year per coop member with the guarantee of delivering 6 kg a week of fresh organic vegetables. They cultivate seventy-five different types of produce, including fruit, vegetables and cereals. The cereals are traditional varieties coordinated in a project funded by the region and supported by the University of Bologna. Finished products include flour made from their organic wheat, and many varieties of pasta.

The farm is located in a protected area called the Rural City Park (Parco Città Campagna), supported by the region and five municipalities. The forty-seven hectares of farmland is owned by one of the municipalities, nestled within a protected area of eighty square kilometers that safeguards the relationship between the city and the countryside

Alternative value chains 101

while developing the agricultural potential of the land. The project also involves community gardens available to a local association of senior citizens.

Similar models on a smaller scale are found in Siena and Rome. In Siena, the Ortomangione project makes use of about 1.5 hectares of land in a peri-urban area which had been abandoned for years. Eighty coop members support the project, reaping the benefits of fresh produce on land they contribute to maintain. They pay an annual quota of €800 which entitles them to weekly produce harvested at the farm. In Rome we find the Semi di Comunità ('Seeds of the community') where members work the land along with a paid staff of three professional farmers. The cooperative yields enough produce to distribute 160 shares of weekly food supplies among as many coop members. The shares are sold at an auction held every February, with a base price determined by total estimated costs divided by 160 shares. They cultivate five hectares of land in a protected nature reserve.

The idea of creating complex locally based value chains finds expression all over the country. One of the least densely populated areas of Italy is the province of Viterbo just north of Rome. It boasts an important experience called The Diffuse Rural Community (CRD) involving various experiences in the short value chain, based on the concept of Zero Km food operations ensuring that products travel no more than 100 kilometers.[28] In Viterbo, the CRD coordinates an ample range of products falling within their network. CRD is the outcome of a process which started in 2019 with a conference organized by a network of organic farmers and activists called 'To eat or to be eaten,' itself part of a movement which in 2010 became 'Wholesome and clandestine.'[29] Their manifesto speaks of a territorial community with a democratic organization based on citizens' assemblies. They seek new ways of linking urban and rural environments, while rejecting GMOs, chemical products, wasteful uses of water and the exploitation of labor. They press for food sovereignty, giving people the right to healthy food at accessible prices, produced on the land where people reside. Land should be treated as a common good, and agriculture should be treated as an instrument of political action. They strive to go beyond the concept of producer and consumer, developing rural and urban commons based on self-organization, solidarity, cooperation and concern for the territory.

The Diffuse Rural Community brings together twenty producers in an area contained within a radius of thirty kilometers. The farms included in the network represent a comprehensive range of products, from fresh produce to olive oils and different types of seed oils, to goat and cow cheese, pulses, legumes, herbs and spices. They have winemakers and craft brewers, people who sell herbs and spices, including locally grown saffron, natural cosmetics and soaps. One producer supplies cured pork products, another makes marmalades. One producer is specialized in autochthonous cereals and makes bread and pasta. Distribution is organized in monthly markets held at four locations scattered throughout the territory as well as a permanent shop in Orvieto. They also sell their products through two local GAS networks operating two local collection points. The network covers a host of needs while delivering food in an ethically and environmentally sustainable framework.

Value chain experiments involving wheat

The two most ancient agricultural products in Italy are wine and wheat. We know that vines play an important role in social agriculture since they have high potential for added value. There is enormous variety in vines and the techniques available to process grapes into wine, and the market is highly competitive. The most rigorous producers advocate a return to the territory, with attention to soil types and plant selection methods, developing micro-territorial products of unique value (Scienza et al., 2019). Attention to experimentation has generated a varied market, including high-quality organic and biodynamic wines produced with attention to social and environmental outcomes.

Pasta is in some ways like wine in that it cannot be consumed without complex processing. The equipment required to make wine is costly, making it hard for small producers to control the value chain process. Wheat processing is instead more easily within in the reach of the small-scale producer. A significant difference between wine and wheat is institutional support. The market for wines links viticulture to the territory, as seen in the abundance of protected wines: Italy has no fewer than 526 protected designation of origin (PDO) wines. Yet it has only one PDO wheat.

Fresh pasta made with eggs is something ordinary families produce with no difficulty. Dry pasta instead requires industrial processing. A critical distinction is the contrast between soft wheat and hard wheat. In environmental terms this corresponds broadly to the climatic difference between continental and Mediterranean areas, corresponding in turn to the distinction between the northern and the central-southern parts of the country.

The most important pasta-making territory of Italy is Gragnano, a well-documented example of an integrated micro-territory which began to produce pasta on an industrial scale in the 1830s.[30] The territory established itself as the world's premier pasta producer owing to the presence of abundant streams available to power mills, proximity to hard wheat producing territories, and the vicinity of major urban markets as well as Italy's first modern port at Castellammare di Stabia. Easy access to the port allowed the area to export pasta all over the world, known by the mid-nineteenth century as white gold. The wheat used for pasta production was originally a variety called Saragolla produced in Puglia, 120 kilometers away from Gragnano. In the nineteenth century, this wheat was replaced by imports from the Ukraine (Falcioni, 2007). The early process was strictly artisan, with pasta being kneaded by foot up to the 1830s. Later the scale of production increased, and more hygienic techniques were introduced. By the turn of the nineteenth century industrial flour mills had developed, a more cost-effective technique than the previous stone milling. This and other technological advances in pasta making heralded the rise of pasta factories all over the world. Pasta workers in Gragnano were rightly concerned about the consequences of these industrial developments. The deterritorialization of pasta making is such that today the leading Italian pasta producer is not the south, with its abundant hard wheat fields, but the industrial north. Indeed, the largest pasta producer in the world is in Parma, the Barilla company founded in the 1870s, precisely the period when roller milling was becoming so common throughout Europe. Yet notwithstanding the advantages of well-funded northern food industries, Gragnano still retains its advantage thanks to the knowledge, skill and reputation that have emerged over many centuries. It is the only PDO pasta in Italy today.

One of the secrets of Gragnano pasta is slow drying, once only sun dried, taking a month to dry in the winter, ten days in the summer. Pasta was so important in this town that zoning laws limited building height to reduce obstruction to the passage of light to street-level drying racks.[31] In industrial pasta production today very high temperature (VHT) techniques are used to dry pasta in a couple of hours, destroying much of the nutritional value of the final product. PDO regulations today specify that Gragnano pasta cannot be dried using temperatures greater than 85°C in a process which must last at least four hours.[32] Another important specification is the use of hard wheat, without however specifying the provenance or the wheat variety. This is the new frontier in pasta making, recognizing that different wheat varieties grown in different territories have appreciable distinction which needs to be given more attention.

An instructive example of extreme deterritorialization is seen in the case of the company Buitoni, founded in Sansepolcro, Tuscany, in the 1820s. The firm was bought by Nestlé in the 1980s, which then sold or closed most of their Italian operations. Nestlé continued to use the historic Tuscan location for the activities of Buitoni's research unit, which among other things develops pasta sauces sold under their brand name and makes recipe recommendations to customers. Then, in 2018, Nestlé closed the Tuscan research facility and relocated it to Solon, Ohio![33] These changes had dramatic consequences for local employment, including the regional agricultural market which recorded a precipitous drop in wheat prices.[34] The consequences for the landscape are yet to be seen.

Global wheat seed production is dominated by multinational companies like Monsanto.[35] Weak awareness of the territorial linkages involved in wheat production is surprising in a country like Italy which has the greatest number of PDO food products in Europe. While the situation is starting to change thanks to initiatives taken by advocates of alternative food networks, for now Italy continues to increase its flour imports, and today half of all wheat consumed or processed in Italy is imported.[36] Yet there are signs that the market is changing. For instance, much pasta sold in Italy today bears the claim that it uses 100% Italian wheat. There are of course exceptions, and Buitoni in 2022 made no mention of the origin of their wheat. They did, however, label their product with the misleading indication that they had been making pasta in Sansepolcro, Tuscany, since 1827. They currently have no presence in that region. Buyer beware!

The wheat market is perplexing, even for producers, In Tuscany, in 2021, the price for organic wheat was lower than for conventional wheat, even though the latter has twice the yield. Lower yields are such that organic food products are only viable economically if they command higher prices than conventional agricultural products. This incongruous market performance is a sign of the weak appeal of organic wheat, partly owing to a lack of familiarity with the autochthonous wheat varieties used in organic production. This weak public awareness means that autochthonous wheat producers must really struggle to maintain commercial viability.

Owing to a lack of sensitivity for the territorial dimension of wheat production and the consequent privileging of cheap imports, even for conventional hard wheat the market price is so low as to make its cultivation in Italy barely sustainable. This is a major concern in many parts of Italy, not least southern Tuscany where wheat has been produced since early Neolithic times. The oldest wheat producing region of the peninsula is the Val d'Orcia, where the harmonious use of the territory over millennia has endowed the landscape with the extraordinary quality which justified its inclusion in the list of UN World Heritage Sites. Yet, the persistence of this timeless heritage is at risk in the modern period. Low market prices and high overheads are driving the stewards of the landscape out of business. The consequences for the territory are cataclysmic, including employment loss. This trend has pushed rural entrepreneurs to seek new production models, striving to control the full farm-to-fork value generation process. They have partly achieved this by developing an industry of holiday farms (*agriturismo*) offering hospitality and meals to a paying public. To qualify as a holiday farm at least half of their revenues must be derived from farming, a condition which ensures support for agriculture. The sale of meals and agricultural products, like cheeses, wine, flour and pasta, allows the farm to generate the revenue they need to be financially sustainable. Tuscany is the leading region for the presence of holiday farms in Italy, an emblem of attention to multifunctional farming outcomes.[37]

In the wine market, there is growing appreciation for Italian autochthonous grape varieties (D'Agata, 2014). This same appreciation is beginning to emerge in wheat markets. Autochthonous wheat varieties are usually publicized as ancient cereals (*grani antichi*). They typically have higher protein content than conventional wheats, as in the case of the autochthonous Solina soft wheat grown in central Italy which has as much as 18% protein. They also have more trace elements and lower gluten content which producers contend is better for one's health. Consumption of food with high gluten content prompts a sugar spike which with time can lead to diabetes. The earliest wheat variety cultivated in Italy is *Triticum monococcum* and has only 6% gluten. Some of these varieties are hulled grains which require special processing to hull the grains before they can be milled into flour. This laborious requirement was once such that hulled wheats like spelt and *T. monococcum* fell out of favor. Today special milling equipment is available at a cost-effective price, and this has led to a revival of some of these ancient wheat varieties.[38] The requirement to process wheat into flour and then flour into pasta is such that ancient wheat growers must be part of a district which includes the presence of specialized millers and pasta makers.

Ancient wheat producers generally purchase seeds from one of many Italian seed companies, usually a single variety determined by local micro-environmental capacity and the needs of the producer. Some of these wheat varieties have acquired public recognition, like

Senator Cappelli which was so important in Italian wheat production in the 1920s. Many producers are returning to this variety because of its close association with the Italian peninsula, and its excellent qualities for pasta making. Another variety which is enjoying success is Timilia, grown in Sicily since the time of Greek colonization.[39] The greatest number of autochthonous wheat varieties in Italy is indeed found in Sicily, where some fifty-two wheat cultivars have been preserved largely thanks to the work of The Experimental Consortium of Wheat Farmers, created in 1927 to preserve local ancient cereals.[40] Sicily is also home to the first regionally protected bread to specify the use of local ancient wheat varieties, the *Pagnotta di Dittaino* made in central Sicily.[41] With time, we may find more food products protected under national PDO legislation which specify local autochthonous varieties, exactly as we have in wine markets. A candidate for PDO recognition is the black bread from Castelvetrano in southern Sicily. It is made exclusively from Timilia.[42]

Other regions are also looking to recapture ancient linkages with specific wheat varieties, like the Abruzzo where Saragolla is enjoying a significant revival thanks also to an EU-funded project aiming to preserve local seed heritage while seeking synergies suited to the geomorphology and climate of Italy's various micro-territories.[43]

An alternative to choosing a single variety is to use a mixture of varieties, the most remarkable being the mixtures of evolutionary populations assembled by the Italian agronomist Salvatore Ceccarelli (2016). He sells packages containing 2,000 varieties of soft wheat, or alternatively 700 varieties of hard wheat, collected all over the world in a lifetime dedicated to preserving the diversity of wheat varieties still available on the planet. The producer sows the complete mixture, and the varieties best suited to a particular micro-territory reach the harvest.

Awareness of micro-territorial characteristics can help develop local productive systems, leveraging multifunctional outcomes and linking the natural potential of the landscape to human capacities developed over generations. Sensitivity to the value chain with strong attention to the territory is what this is all about. The Italian territorialists have used this approach with a significant participatory component built into a bioregional framework (Magnaghi, 2005, 2020). We have an excellent food-related example of their work in the area of Prato, a city with an important manufacturing history on the Arno River plain to the west of Florence. This is the Gran Prato project, a low-cost grassroots initiative supported by local institutions and by the University of Florence. Gran Prato is the name of a local bread which has been

Alternative value chains 107

developed as a socially and environmentally embedded value chain aiming to overcome the food desert which separates producers from consumers. This project strengthens rural–urban linkages by connecting stretches of the Arno floodplain to the adjacent residential areas. Major rivers are always a challenge from the standpoint of landscape management because they flow through territories under separate administrations. The Arno flows through fifty townships each having primary authority for the management of their respective territories, organized within the framework established by the regional integrated territorial plan (PIT). The coordination of conservation and development efforts is also provided by a 'river contract.'[44]

The Arno floodplain has been heavily fragmented by a long history of industrial development (Poli, 2016). The Gran Prato project aims to promote integration by raising public awareness concerning the importance of the riverine landscape using a local food product. This is a traditional bread using local wheat, local transformation systems, local retail networks and local end consumers. Some of the farming areas involved in the project are threatened by real estate development and have already been parceled out for residential construction even though still not zoned for this purpose (Poli, 2016:81). The Gran Prato Association was created in 2013 to coordinate the project, bringing together local cereal farmers, millers and baker-retailers. The wheat is organic and must be transformed into bread within four to five months of being milled.[45] Information concerning the initiative is circulated by word of mouth, publicized in the retail outlets, and disseminated through information services provided by the City of Prato. The website Eatprato lists this bread along with other products which express concern for the local territory that goes beyond mere food provision.

This integrated local system is a small but important step in the complex process of protecting the landscape by building connections between farmers and consumers. The project uses the short value chain to strengthen a biodistrict comprising coordinated linkages between producers and consumers, urban metabolism and the natural bounty of the land. The initiative shows how multiple stakeholders can be involved in the stewardship of the territory. The concept of the biodistrict helps re-territorialize food with benefits for climate change resilience, environmental sustainability, interclass social cohesion, food co-production and co-governance, food justice, landscape quality and the circular economy (Poli, 2017).

The Arno riverside agricultural park reintroduces the issue of healthy and local food production in planning strategies and actions, while responding to a new peri-urban rurality and satisfying

108 *Alternative value chains*

the emerging critical needs of a highly urbanized setting. 'Farming with the Arno' triggered among the partners of the project awareness of multifunctional peri-urban agriculture and food value chains, with attention to issues of governance and landscape planning (Poli and Butelli, 2021). The project is one of the ongoing instruments adopted in the 2030 Strategic Plan of the Metropolitan City of Florence.

Notes

1 For a contemporary design approach to the neighborhood unit concept, see EVstudio 'The neighborhood unit: how does Perry's concept apply to modern day planning?' https://evstudio.com/the-neighborhood-unit-how-does-perrys-concept-apply-to-modern-day-planning/ (Accessed August 5, 2022).
2 Gregory Smith carried out an investigation of this neighborhood with his Cornell University Rome Workshop students in the spring of 2018. The student report can be found at https://aap.cornell.edu/sites/default/files/upload/Torre%20Maura%20Final%20Report-VD.pdf (Accessed July 14, 2022).
3 Interview with Francesco Paniè of Terra! Onlus in the spring of 2018.
4 https://www.unioncamere.gov.it/P42A4251C160S123/-italia—itinerante—oltre-180mila-operatori-nel-commercio-ambulante—il-56—parla-straniero-ma-in-alcuni-comuni-i-banchi-italiani-sono-oltre-il-90-.htm (Accessed July 14, 2022).
5 https://www.controcampus.it/2021/03/diventare-fruttivendolo-corsi-stipendio-cosa-fa-in-giornata/ (Accessed July 14, 2022).
6 https://www.comune.napoli.it/flex/cm/pages/ServeBLOB.php/L/IT/IDPagina/9640 (Accessed July 14, 2022).
7 https://dati.comune.milano.it/dataset/ds291-economia-mercati-settimanali-scoperti/resource/5d4a83c3-6ac5-4430-b5a9-d669a933b036 (Accessed July 14, 2022).
8 https://www.comune.fi.it/pagina/imprese-e-commercio/mercati-rionali (Accessed July 14, 2022).
9 https://www.comune.venezia.it/it/content/mercati-e-fiere (Accessed July 14, 2022).
10 https://www.romatoday.it/economia/ddl-concorrenza-ambulanti-roma-bolkestein-raggi.html (Accessed July 14, 2022).
11 https://storiedimenticate.it/il-primo-supermercato-ditalia/ (Accessed July 14, 2022).
12 https://storiedimenticate.it/il-primo-supermercato-ditalia/ (Accessed July 14, 2022).
13 https://www.e-coop.it/comunicati (Accessed December 32, 2021).
14 https://www.beverfood.com/responsabilita-sociale-impresa-coop-migliore-catena-grande-distribuzione-europa-wd3346/ (Accessed December 23, 2021).
15 https://www.associazioneterra.it/2021/11/04/aste-al-doppio-ribasso-vietate/ (Accessed July 14, 2022).
16 https://www.sinlabor.it/2018/08/24/pomodoro-rosso-ipocrisia/ (Accessed July 14, 2022).

Alternative value chains 109

17 https://www.ideegreen.it/farmers-market-114987.html (Accessed July 14, 2022).
18 https://www.coldiretti.it/vendita-diretta/11-acquisti-vola-la-spesa-campagna (Accessed December 15, 2021).
19 https://www.coldiretti.it/vendita-diretta/11-acquisti-vola-la-spesa-campagna (Accessed December 15, 2021).
20 Paxton, Angela (1994) *The food miles report: the dangers of long-distance food transport.* London: SAFE Alliance. http://www.sustainweb.org/publications/?id=191 (Accessed December 15, 2021).
21 https://www.treccani.it/magazine/lingua_italiana/articoli/parole/chilometri_zero.html (Accessed December 15, 2021).
22 https://www.designmakers.it/interior-design/mini-fabbriche-roma-il-design-e-a-km-0/. https://www.bottegachilometrizero.com/philosophy/?lang=en&v=cd32106bcb6d (Accessed December 28, 2021).
23 http://www.economiasolidale.net/sites/default/files/allegati/2007_CartaRES.pdf (Accessed December 28, 2021).
24 https://e-circles.org/mappa-gruppi-di-acquisto (Accessed December 28, 2021).
25 https://osservatoriocoesionesociale.eu/osservatorio/il-consumo-responsabile-in-italia-i-primi-dati-dellindagine-2020/ (Accessed December 28, 2021).
26 https://abruzzoweb.it/legge-sui-gas-i-beneficiari-dicono-no-non-ci-servono-soldi-pensate-ad-altro/ (Accessed December 28, 2021).
27 https://humusjob.it/magazine/rete/le-comunita-a-sostegno-dellagricoltura/ (Accessed December 28, 2021).
28 https://comunitaruralediffusa.org/il-percorso/?v=07feabf0a4cf (Accessed December 28, 2021).
29 https://genuinoclandestino.it/chisiamo/ (Accessed December 28, 2021).
30 https://www.consorziogragnanocittadellapasta.it/storia-del-prodotto/ (Accessed December29, 2021).
31 http://www.agricoltura.regione.campania.it/tipici/disciplinari/Disciplinare_Pasta_di_Gragnano_12-02-2020.pdf (Accessed December 10, 2021).
32 https://ilfattoalimentare.it/pasta-spaghetti-essiccazione-temperatura.html (Accessed December 10, 2021).
33 https://www.repubblica.it/economia/2018/04/02/news/buitoni_centro_ricerca_ohio-192810926/ (Accessed December 10, 2021).
34 https://www.umbria24.it/economia/grano-duro-crollo-umbria-del-30-15-tenero-spariscono-girasoli (Accessed December 10, 2021).
35 https://altreconomia.it/i-custodi-dei-semi/ (Accessed December 12, 2021).
36 https://www.ilsole24ore.com/art/pasta-il-grano-made-italy-basta-appena-quattro-mesi-all-anno-ACz8v9U (Accessed December12, 2021).
37 https://www.istat.it/it/files//2020/11/Le-aziende-agrituristiche-in-Italia-2019.pdf (Accessed December12, 2021).
38 https://wholegrainscouncil.org/blog/2014/09/separating-wheat-chaff#:~:text=This%20can%20require%20two%20processes,the%20hull%20is%20already%20loose (Accessed July 22, 2022).
39 http://www.gentedelfud.it/prodotto/dettaglio/grano-timilia-o-tumminia/ (Accessed July 22, 2022).
40 Stazione consorziale sperimentale di granicoltura. https://www.crea.gov.it/documents/20126/0/Manifesto_Biodiversity_day_2019_02.pdf/

a7ba6bf2-e960-12e9-18f2-fc86d667c9ff?t=1559296177654 (Accessed July 22, 2022).
41 https://www.politicheagricole.it/flex/cm/pages/ServeBLOB.php/L/IT/IDPagina/16066 (Accessed July 14, 2022).
42 http://www.fondazioneslowfood.it/presidi-italia/dettaglio/3560/pane-nero-di-castelvetrano (Accessed January 2, 2022).
43 https://www.innovarurale.it/it/pei-agri/gruppi-operativi/bancadati-gopei/progetto-operativo-di-innovazione-i-grani-antichi (Accessed January 2, 2022).
44 River contracts implement EU and national legislation in the management of waterways. https://www.appenninosettentrionale.it/itc/?page_id=9716 (Accessed January 2, 2022).
45 http://www.eatprato.it/it/il-territorio/ (Accessed January 2, 2022).

Bibliography

Black, Rachel E. (2014) *Porta Palazzo. The anthropology of an Italian market.* Philadelphia: University of Pennsylvania Press.

Boccia, Flavio and Daniela Covino (2018) *Corporate and consumer social responsibility in the Italian food market system.* London: Intech Open. https://doi.org/10.5772/intechopen.75309.

Cattaneo, Carlo (1965 [1859]) *Scritti politici.* Edited by Mario Boneschi. Florence: Le Monnier.

Ceccarelli, Salvatore (2016) *Mescolate contadini, mescolate. Cos'è e come si fa il miglioramento genetico partecipativo.* Savona: Pentagora.

Clancy, Kate and Ruhf Kathryn (2010) 'Is local enough? Some arguments for regional food systems,' *Choices*, 25:1: 1–5.

Corrado, Alessandra (2011) 'Clandestini in the orange towns: migrations and racisms in Calabria's agriculture,' *Race/ethnicity: multidisciplinary global contexts*, 4:2:191–201.

D'Agata, Ian (2014) *Native wine grapes of Italy.* Oakland: University of California Press.

Dansero, Egidio and Matteo Puttilli (2014) 'Multiple territorialities of alternative food networks: six cases from Piedmont, Italy,' *Local environment*, 19:6:626–43.

Falcioni, Paola (2007) 'Vocazioni ambientali e tradizioni culturali nella produzione della pasta. Gragnano e Fara San Martino,' *Geostorie. bollettino e notiziario del centro italiano per gli studi storico-geografici*, 15:1:23–39.

FIVA (Federazione Italiana Venditori Ambulanti) (2017) *I mercati ambulanti. Un'indagine sul vissuto e sulle potenzialità di sviluppo.* Milan: Euromedia Research.

Knights, Melanie (1996) 'Bangladeshi immigrants in Italy: from geopolitics to micropolitics Rome,' *Transactions of the Institute of British Geographers*, 21:1:105–23.

Magnaghi, Alberto (2005) *The urban village: a charter for democracy and local self-sustainable development.* London and New York: Zed Books.

Magnaghi, Alberto (2020) *Il principio territoriale*. Turin: Bollati Boringhieri.
Manzini, Ezio (2021) *Abitare la prossimità. Idee per la città dei 15 minuti*. Milan: Egea.
McCorriston, Steve (2006) 'Trade liberalization under imperfect competition in commodity markets,' in *Governance, coordination and distribution along commodity value chains*, Proceedings of the FAO Commodities and Trade. Rome: FAO, pp. 27–46.
Montanari, Massimo (2008) *Il formaggio con le pere. La storia di un proverbio*. Bari: Editori Laterza.
Perry, Clarence A. (1929) *The neighborhood unit*. New York: Regional Plan of New York and Its Environs, Vol. VII.
Petrini, Carlo (2005) *Buono, pulito e giusto. Principi di una nuova gastronomia*. Turin: Einaudi.
Poli, Daniela (2016) 'Rivoluzione alimentare e parchi agricoli multifunzionali nella piana fiorentina,' *Economia e società*, 14:2:70–91.
Poli, Daniela (2017) 'Food revolution and agro-urban public space in the European bioregional city,' *Agroecology and sustainable food systems*, 41:8:965–87.
Poli, Daniela and Elisa Butelli (2021) 'Una nuova ruralità periurbana nel cuore della città metropolitana: un parco agricolo multifunzionale in Riva sinistra d'Arno,' *CRIOS. Critica degli ordinamenti spaziali*, 22:30–43.
Russo, Michelangelo and Arjan van Timmeren (2022) 'Dimensions of circularity for healthy metabolisms and spaces,' in Libera Amenta, Michelangelo Russo and Arjan van Timmeren (eds.) *Regenerative territories: dimensions of circularity for healthy metabolisms*. New York: Springer Geojournal Library, vol. 128, pp. 130–27.
Scienza, Attilio, Alessandra Giorgianni, Diego Tomasi, Federica Gaiotti, Federico Graziani, Luigi Mariani and Paolo Carnevali (2019) *Atlante geologico dei vini d'Italia: vitigno, suolo e fattori climatici*. Florence: Giunti Editore.
Smith, Gregory (2021) *Urban narratives and the spaces of Rome: Pier Paolo Pasolini and the city*. Abingdon: Routledge.
Soja, Edward W. (2011) 'Regional urbanization and the end of the metropolis era,' in Gary Bridge and Sofie Watson (eds.) *New companion to the city*. Chichester: Wiley-Blackwell, pp. 679–89.
Sonnino, Roberta and Terry Marsden (2006) 'Beyond the divide: rethinking relationships between alternative and conventional food networks in Europe,' *Journal of economic geography*, 6:2:181–99.

6 Public administration and innovative value chain management

Four factors should govern the post-COVID-19 administration of the territory: environmental sustainability, economic development, equity and citizen participation. In order to achieve these goals policy actions should operate at two levels. One concerns public policy focusing on the broader workings of the agri-food system, the other should focus on individual choices. Landscape management in the context of agroecological systems is one dimension of the dilemma, linked to a contextualized framework of individual choice and personal health. These two dimensions – the shared and the personal – are strictly interconnected. In times of crisis for public finances and the corresponding incapacity of the state to provide essential services, the search for alternatives to neoliberal policy models is ever more compelling. More support must be provided for alternative models of economic development which are alert to the multiple outcomes of food-related activities. The alternative policy must also support the capacity to generate prosperity across the community. This is the mandate provided by UN Habitat to create stronger urban–rural continuity, sharing prosperity while paying greater attention to environmental sustainability (UN Habitat, 2019). The plan stresses the need to promote economic growth while improving the environment and creating effective urban crisis prevention and response.

Agri-food systems are unique in their linkage of individual choices to shared outcomes. As Wendell Berry claimed, eating is an agricultural act, and what people eat has direct impact on farm choices and opportunities, which in turn impact the quality of our shared environment. The aesthetic dimension is important: a rich natural landscape can shape rich personal identity. The landscape not only reflects human aesthetic standards but also must be respected on its own terms, independent of human interests and activities. The human presence

Public administration and innovative value chain management 113

in nature expresses a two-way process: the reciprocal permeability of people and the world around them (Iovino, 2016). The reciprocal interdependence between humankind and nature cannot be overstressed, bearing in mind that nature can exist without humans, but humans cannot exist without nature (Latour, 2017). A fundamental concept in maintaining a balanced relationship between the two is critical consumption, acknowledging the plural consequences of our actions in an economic process going beyond the satisfaction of individual desire. A significant consequence concerns the landscape which the Italian constitution rightly protects as an expression of shared heritage. Many studies have chronicled changes in the social ecological system, eliciting the problem of how to evaluate intangible and non-monetary landscape values (e.g., Cerreta and Poli, 2017). Some scholars assert that the landscape is a more suitable concept than the ecosystem when attempting to characterize contributions to individual wellbeing (Vallés-Planells et al., 2014). The landscape martials a social dimension focusing on the spatial interaction of natural and human processes. It is a holistic and dynamic spatial and mental entity resulting from people–place interactions. This interpretation is consistent with the definition of the 2000 Council of Europe European Landscape Convention.[1] The focus on human perception establishes the priority of encouraging citizens to understand the landscape impact of individual actions. This is the foundation for modern EU policy shaping awareness of the deep interpenetration between people and place, inviting citizens to answer for their individual actions.

Digital communities play a role here. Many citizens are in the grips of a media system which fragments the community and splinters the individual, making them more susceptible to the marketing message of a no longer sustainable economic system (Dear and Flusty, 2000). Digital media in particular facilitates this function. We need a new netnographic approach integrating digital and conventional communities, exploring the process of cocreation used to draw together shared consciousness, shared rituals and traditions, and a shared sense of moral responsibility (Costello et al., 2017). Alternative food networks (AFN) do precisely this: they use Internet to strengthen real-life communities, to forge new ideas, to facilitate ongoing activities. Rather than annihilate the physical world, AFN activists use social media to strengthen real life relations embedded in a real-life world.

A shared sense of moral responsibility must reshape the perceptions of ourselves and the communities in which we live. One writer notes a similarity between the conventional corporate mindset and that of mafia-type crime groups which mandate loyalty to the group, obedience and honor.[2] Italy's small-scale fragmented economic system instead encourages more attention to social relationships, concern for outcomes beyond economic advantage. This sensitivity needs to be extended to our relations with the physical world. Anti-mafia activism seen in this light is more than a struggle against mafia formally understood. It is a struggle against self-aggrandizing actions which optimize personal advantage to the detriment of the social and natural environment; the same environment that will determine the ultimate success or failure of any human enterprise. This is what corporate social responsibility is about. Focus on the idea of broad responsibility has received academic support through the idea of 'cooperative competition,' rendered more poetically in Italian as *concuranza* ('shared concern' is a reformulation of the word for competition, *concorrenza*) (Alvisi, 2021). We see this logic in Italian neighborhood markets, where stallholders do not compete on price, knowing that their common destiny is to survive or fail as a community. Here we have a new paradigm for a new world.

This is the foundation of the social economy, supported by Law 381 of 1991 which identifies social cooperatives which in turn make possible important forms of local solidarity. Farming is an activity which can benefit from a variety of skills and cultural backgrounds. The therapeutic applications in the USA and the UK are referred to as care farming, where farming has the stated purpose of pursuing health-related outcomes, as well as social or educational ones. In Italy the importance of social farming in supporting welfare activities is recognized in Law 141 of 2015 which defines the concept of social agriculture with reference to multifunctionality. Farms qualifying for this recognition deliver services relating to social, health and educational outcomes, they provide work opportunities for differently abled citizens, and essential services to people living in a condition of marginality. Authorized social farms are given various forms of support, such as priority in selling their products to public amenities like schools and hospitals, or priority in the assignment of confiscated mafia property. Social farming is expanding in Italy, and whereas most of these farms are in the north some southern regions have documented important experiences in using social farming to support modern welfare needs (Fondazione Campagna Amica, 2020). Some of these farms, like those in the Region of Campania, have been created through the assignment of confiscated mafia property, showing the radical potential of using the reassignment of mafia assets not only against mafia, but more importantly in favor of a truly inclusionary society.

Public administration and innovative value chain management 115

In the modern interconnected world with its increasing urban populations, linkages between farming communities and urban populations have received increasing attention. Indeed, cities are the drivers in the global food system, including its many negative externalities (Calori et al., 2017). Their leadership position allows cities to play a virtuous role in promoting more sustainable urban–rural linkage. A key framework for driving positive change is the Urban Food Policy (UFP), city-based plans which deal with the global problems of climate change, food insecurity and food-related conflicts (Morgan, 2015). UFPs create sustainable, resilient and fair food systems, with new spaces of deliberation involving multiple actors. Urban food planning discourse has solid North American and British roots and has found increasingly fertile soil in Italy. Pisa in 2010 was the first Italian city to devise a UFP (Di Iacovo et al., 2013). This was a starting point which led up to the Milan Food Policy Pact (MUFPP) of 2015, signed by 138 cities scattered around the globe. The pact provides an important comparative framework for experiments in assessing sustainable agri-food systems that stretch from the farm to the fork and on to the waste bin. Today increasing numbers of cities are devising UFPs. One of the most interesting experiences in urban food policy is seen in the well-documented process in Turin as part of the Torino Metropoli 2025 plan (Di Bella, 2016; Calori et al., 2017). The Turin process involves a wide range of actors, including municipal government, but has yet to be formalized as a UFP (Calori et al., 2017:1041). The Milan project was instead more focused, received strong support from the municipal government, and led effectively to the drafting of an UFP. In both processes, the participatory dimension was of paramount importance.

An assessment of the UFP strategies adopted in Milan and Turin has led to various observations (Calori et al., 2017). The first is that UFPs should be interconnected among multiple urban centers rather than focusing on one alone, articulating all the scales required to engage the relevant levels of territorial authority. The second is that actors within the system should be left with substantial autonomy rather than be compelled to fit into a predetermined scheme. A third observation, looking specifically at the Italian experience, is that the process of deterritorialization is less advanced here than it is in North America and the UK.

Given Italy's strong historical emphasis on place, it may be easier here than elsewhere to revive sensitivity to the territorial dimension of the agri-food system. This revival is assisted by persisting fragmentation, facilitating a new yet old equation revolving around awareness of food's multifunctional character. Italian rurban linkages are especially strong at the peri-urban scale, precisely the level of rich experimentation all over the world in the spatial re-embedding of food chains (Feagan, 2007). Italian research has shown that this peri-urban dimension can be further subdivided in two different scales – one of proximity and the other of the bioregion (Poli, 2017). The city of proximity is the everyday city, where through minimal movement citizens can buy directly from farms or farmers markets, come into direct contact with the world of agricultural production. It is the edible city, where food is an opportunity for socialization and the redevelopment of living spaces. The bioregion is instead the foodshed understood as a socio-geographical space. The bioregion goes beyond individual action, a larger integrated community involving all the different phases in the production and distribution of food. The city of proximity is the city of personal choice, while the bioregion requires a more complex interaction of social, natural and institutional forces. Increasing attention to the bioregion is prompted by escalating public health costs, impelling threats to natural resources, and challenges faced by public security. These concerns put public food policy at the center of contemporary political debate assessing the challenges of multilevel governance.

Milan Urban Food Policy Pact

The Milan World Expo of 2015 organized under the title 'Feeding the planet, energy for life' was the continuation of a process of integrating rural and urban communities in Italy's industrial north that started in the early millennial years. The Expo provided the occasion to launch an international initiative promoting urban food policy plans all over the world under the heading of Milan Urban Food Policy Pact (MUFPP). The Expo gave Italy significant visibility, drawing together 145 countries. Milan was a particularly apt host for the initiative, being one of Europe's most important metropolises as well as home to the largest peri-urban agricultural park in Europe. This is the Agricultural Park of South Milan situated in the confines of sixty-one different townships, covering forty-six square kilometers. It constitutes an excellent showcase for the sustainable management of linkages between urban and rural populations. The process which eventually became Milan's UFP started in 2002 with the upscaling of grassroots GAS networks, bringing together farms, and other actors committed to principles of fair trade and solidarity economy. The coalition included ethical banks and microcredit actors, forming in 2009

Public administration and innovative value chain management 117

a coalition called the District of Rural and Social Economy (DESR) (Calori et al., 2017). Between 2011 and 2013, four agricultural districts were created in the metropolitan area through a lengthy participatory process, coordinating active farmers and local authorities in a formal framework. This socially driven process led to the approval in 2015 of the Guidelines for the Milan Food Policy 2015–2000 to improve the efficiency of Milan's food system (Resolution 25, October 15, 2015). The five priorities of the plan were access, sustainability, education, waste and research.[3] The resolution indicates that the food plan would entail no cost to the City of Milan, in a process driven mostly by community actors (Calori et al., 2017:1038). This same resolution gave official life to the MUFPP which was signed during the 2015 Expo by 100 cities, and today has 200 signatories.

This complex process was sponsored by Milan's progressive mayor, Giuliano Pisapia, funded by the Cariplo Foundation. The loose initial mandate was to make the city more sustainable by strengthening the relationship with the surrounding rural areas. The first phase of the project involved factfinding and brainstorming shared among fifty scholars and fifty associations in a process that aimed to combine scientific expertise and social knowledge. The second phase was articulated through a series of public meetings involving community actors and business enterprises. The process was intended to give life to a permanent food council to discuss and evaluate food policy activities, implementing measurable targets and indicators (Calori et al., 2017). A major focus was to reduce food waste, making healthy food available to the entire community. This was achieved by creating a network of neighborhood-based facilities, called food hubs, made available by the city and staffed by volunteers. The food hubs are equipped with storage rooms and refrigerated spaces, and carry out distribution operations following procedures monitored by researchers from the Milan Politecnico. Tax incentives encourage supermarkets and cafeterias to participate in the program by donating food. A network of NGOs then redistributes the food collected by the five hubs among the city's needy citizens. This is articulated within the broader framework established by the Milan 2015–2020 Food Policy.

The MUFPP today is an international movement promoting sustainable urban food systems all over the globe. The agenda involves six action categories articulated within six pact categories aiming to find local solutions for global problems. The Pact stresses the role of smallholder producers in the equitable promotion of culturally appropriate food production, using urban and peri-urban agriculture to promote biodiversity and integrate urbanized regional landscapes. The Pact expresses particular concern for food deserts where healthy food is not available especially in low-income neighborhoods.[4] The urban focus will eventually serve the needs of 50% of the current world population, a percentage anticipated to grow to 70% by 2050. The pact categories are governance, sustainable diets and nutrition, social and economic equity, food production, food supply and distribution and food waste. Among the performance indicators for this plan are urban access to community gardens and municipal policies encouraging agricultural production and processing within their territories.

The MUFPP provides a global framework for developing local policies aiming to address the range of issues concerned with food and food security. The Food and Agricultural Organization became a member of the Pact in 2015 and offers important support in helping to measure outcomes of local food policies (FAO, 2019). It is important to note that the MUFPP does not suggest that it would be either possible or desirable to solve food security issues entirely at the local level. But the bottom-up perspective can alleviate the consequences of extreme urban poverty which is estimated to involve over 50 million people in Latin America and the Caribbean alone (Martín and de la Fuente, 2022:10). This estimate was before the Ukraine crisis which is likely to have deeply disrupting consequences for food security all over the world (FAO, 2022). Local food policy can optimize food systems in terms of environmental impact and social inclusion, while preserving biodiversity and distinctive cultural traits. It also helps ensure healthy diets, not only in terms of macro elements like carbohydrates, fats and proteins, but also in terms of so many trace elements which are known to have decisive impact in protecting citizen health (Pollan, 2008). The MUFPP is a remarkable example of a complex multilevel governance process in which the whole city takes part, promoting exchanges with other Italian municipalities in the struggle against food waste. It is also part of European and international cooperation projects (like Urbact and Horizon 2020). The Milan City Council was involved as an advisor by the European Commission in the definition of the urban aspects of European policies concerning food, such as Food 2030, and other food security initiatives (Calori and Federici, 2021).

Turin Food Plan

Turin is an ongoing case of virtuous urban development, starting in 1993 when Italy's car-making capital had to negotiate the conversion of a manufacturing city dominated by one industrial group to a multipolar dynamic center of innovation with strong attention to green issues and food. Today Turin is a major world capital known for quality food projects, with actors like Slow Food and Eataly supporting a host of events and activities involving quality and culturally appropriate food products. Turin is the capital of Piedmont Region, an area noted for outstanding food production and well-managed landscapes, including the Langhe with its superb wine heritage and a landscape inscribed in the UNESCO World Heritage List. Their focus is on innovation through networks allowing a high degree of autonomy among the actors involved.

Starting in the early millennial years, three different processes have generated a series of rich initiatives promoting sustainable and resilient urban food and environmental systems. One is the working table called Turin Food Capital, started in 2014 by a public-sector association called Strategic Turin bringing together the City of Turin, the Chamber of Commerce, two universities and the San Paolo Foundation (Calori et al., 2017). This table aims to create a food commission combining a policy council and a business hub. The second process is Feeding Metropolitan Turin, launched in 2015 as a participatory process managed by what was formerly the province of Turin, then the Metropolitan City of Turin. This participatory project aims to draft a local food agenda and involves the University of Turin along with a wide range of actors in the food system. The third process is the Food Smart Cities for Development, an international project funded starting in 2015 by the European Commission's Development Education and Awareness Raising (DEAR) program. It involves twelve urban areas, including Turin, over three continents coordinating food policy and international cooperation. The shared focus of all three projects is inclusionary access to high-quality food understood in economic, cultural and physical terms. The three processes are currently coordinated by the Food Atlas (Atlante del Cibo) which monitors the food system with participatory methodologies. The Food Atlas is an association bringing together local authorities, a foundation, universities and the business community to coordinate Turin's food system. In 2020, Food Atlas coordinated a memorandum of understanding among the different territorial levels, from the city to the metropolitan authority, to the region. While not constituting an urban food policy in a formal sense, it represents a process which has all the functions of an UFP. Thanks to these processes, Turin now has the most systematically investigated food system of any Italian city, all linked to the UN 2030 Agenda for Sustainable Development Goals. The publications that document this investigation include the Turin Food Plan (TFP; Bottiglieri et al., 2017), a detailed strategic plan for the city's green infrastructure (*Piano strategico infrastruttura verde*, 2020) and a climatic resilience plan (*Piano di resilienza climatica*, 2020).

The slogan for the TFP is 'Eat fair and local, change the world.' Turin is one of the first major cities in the world to insert in their statute the right to adequate food 'understood as sufficient nutritious and healthy quality food acceptable from the cultural and religious standpoint, capable of guaranteeing the mental, physical, individual and collective needs required to conduct a dignified life.'[5] It is also one of the first cities to integrate within a comprehensive food and agriculture framework all the relevant territorial levels. The project brings together multiple cities within a metropolitan framework, coordinating the actions of public authorities with those of universities and associations, and citizens from all walks of life. The strategy creates a local territorial food system integrating civil society, government and the business community. A crosscutting theme is food sovereignty, a goal which requires a high degree of local autonomy in the political, legal, organizational and financial autonomy realms. Food sovereignty also hinges on a decisive grassroots component articulated through participatory democracy. The TFP not only sets out this framework but also provides background on the various strategies which have already been implemented in pursuing this plan. Many of these strategies are inclusionary, such as providing meals for the needy, food support for fragile citizens, programs for environmental education, school meals that are culturally and environmentally sound, school gardens and urban farms with direct sale to the public. The plan also describes urban herders, illustrating how Turin is the first city in Italy to allow livestock to graze in public parks, providing maintenance in exchange for pasture. It also describes the Four Cities for Development, an EU-funded project where four European cities work with Slow Food to support typical food production in developing countries. The City of Turin provides support for Slow Food presidiums in Senegal and Mali, including the Fadiouth Island salted millet couscous presidium.[6] The city also promotes food related ethical tourism in developing countries. The number and range of projects is impressive and sums up all the best practices that can be expected of an UFP.

No less impressive is the detailed Strategic Plan for Green Infrastructure (*Piano strategico infrastruttura verde*, 2020) and its Climatic Resilience Plan (*Piano di resilienza climatica,* 2020). The green infrastructure plan was devised by the city's department for environmental policy and public green as part of the Turin 2030 project called 'Turin Livable City,' one of the four dimensions of Turin 2030 focusing on sustainability and resilience.[7] The plan sets out the context for the study, including the policy framework in which it operates. Important support for this initiative comes from Italian legislation, especially Law 10 of 2013 establishing measures for developing urban green areas. The Italian Association of Municipalities also provided support, in the form of guidelines established in 2017. The process necessarily involved interaction among different levels of administration, from national government to regional and municipal planning bodies.

The Strategic Plan for Green Infrastructure sets out the all-embracing role of public green, noting the multiple functions of the natural environment: social, ecological, cultural and economic. A strong green system encourages a healthy lifestyle, embedded in meaningful social relations which further community integration. The green system is fundamental in improving the urban microclimate while mitigating pollution and climate change. The plan supports biodiversity and promotes the fundamental role of the natural landscape in the process of urban regeneration, no small issue in what is potentially a rust belt city.

The green infrastructure comprises a range of spaces, from neighborhood gardens to every scale of urban park, including those in the hills around the city, and the riverine parks lining the four rivers that cut through the urban territory. It comprises professional farms and the all-important forests, listing monumental trees of special botanical and landscape interest. Many spaces are recreational green areas, classified according to their proximity to individual residences. Over 90% of city residents have a recreational park within 300 meters of their homes. These parks are equipped with fitness areas, barbecues, assembly spaces and even musical instruments. Community gardens receive special attention in the study, noting that about 75% of these gardens are spontaneous, having come into being with no public sector involvement. The other community gardens belong to associations, to schools, or are organized by the local boroughs. The plan is remarkable for its detail, documenting the current situation and proposing blueprints for future development.

The climatic resilience plan is similarly comprehensive, illustrating environmental strategies often linked to natural remediation. The plan is articulated within the EU climate change framework, in particular the 2007 Green Book which led to the *European Strategy for Adapting to Climate Change* of 2013. This in turn led to Italy's 2015 *National Strategy for Adapting to Climate Change*, which mandated the plan established by the Piedmont Region in 2017. The regional plan is based on the twin pillars of adaptation and resilience. In this connection, Turin drew up its own climatic resilience plan adopted in 2019. The plan starts by mapping out the territories with the greatest risks, including the Mediterranean basin, the Alps, the floodplains and heavily populated areas. Flooding is a particularly important risk factor in a city like Turin, traversed by major rivers. Having mapped out the risk factors, the resilience plan then assesses the costs and benefits of competing adaptation strategies. Temperature extremes, for instance, are a significant threat in dense urban environments, and can be dealt with using nature-based solutions to foster resilience in urban heat islands. This is achieved by protecting and extending the city's extensive green infrastructure.

Other Italian cities

Many Italian cities signed on to the MUFPP but in fact have done little to devise and implement a UFP. Rome's signed on to the MUFPP in 2015, under its progressive mayor Ignazio Marino, who was ousted from office in the same year and had no opportunity to implement a plan. We recall that Marino was responsible for drawing up the regulations for community gardens after decades of uncertainty. Since his tenure as mayor, Rome has experienced shifts in political leadership which have made it impossible to pursue the creation of a UPF plan in a linear fashion. In the 2016 elections Virginia Raggi of the Five Star Movement won office and began work on what was called the AGRIFOOD Roma 2030 initiative. This was intended to furnish the foundation for the sustainable development of the agri-food sector, articulated in a value chain perspective anchored to specific features of the territory. The plan describes the importance of Rome's neighborhood markets as facilities concerned not only with food but also with social, education and cultural activities.[8] The campagna romana and its typical products are not only developed and protected as features of the local food system, but also as a tourist resource for Italy's most important tourist destination. The initiative aimed to create stronger linkages between Rome's 144 neighborhood markets and the local production system, encouraging public awareness of important rural–urban ties. Throughout concern is expressed to reduce waste and build a more robust system of ensuring access to healthy food for all citizens.

Many stakeholders were involved in exploring the strategic dimensions of AGRIFOOD Roma 2030, but no document was produced before the 2021 elections brought into office a mayor running on the Democratic Party ticket, Roberto Gualtieri. The first two priorities in his campaign were the fifteen-minute city and the green city, creating more supportive neighborhoods nestled in a more sustainable environmental setting. In this connection his plan also speaks of developing local level agriculture.⁹ Within the first few months in office Gualtieri began work on a new URP plan for Rome, establishing once more a series of round tables to assure a participatory process. As with the previous plan, a wide array of local stakeholders was involved, many of whom had been active in the previous participatory framework. Some members noted how the change in local government represented a form of discontinuous continuity. Many actors from the earlier process were also involved in the later one, pursuing similar initiatives in the guise of a completely new framework.

In February 2022, the City of Rome announced the creation of a Food Council which would furnish indications on how to draw up Rome's UFP, following a city ordinance from the previous year which had already announced this intention. The 2022 plan involved the creation of seven round tables involving all the stakeholders in Rome's agri-food system, divided according to the different strategic spheres involved in the URP.¹⁰ The first table is concerned with food governance, the problem of how to coordinate the leadership of the different interests involved in the agri-food system. The second table is concerned with local agricultural resources and what is termed agroecology. This table aims to involve young people in agriculture and establish new farms on public land as a way to reduce urban sprawl and protect the landscape. The third table concerns food logistics and local markets, ensuring the protection of legitimate local products and the exclusion of those created through such illegal means as labor exploitation. The fourth table is organized around the creation of Districts for the Solidarity Economy, including solidarity purchasing groups and AFN, as well as the promotion of social agriculture and community gardens. The fifth table is concerned with waste and inclusionary policies aiming to ensure that all citizens have access to healthy food. The sixth table is concerned with school meal plans, and green procurement for all public sector activities ensuring that in all food related activities priority is given to local products. The seventh table is concerned with gastronomic culture, including restaurants, helping add value to local food products and create greater awareness of excellence in the local agricultural sphere.

As of 2020 some thirty Italian municipalities were working on UFPs (Marino et al., 2020), involving a host of actors, including the Cities of Wine association representing Italy's leading wine producing municipalities.¹¹ Bari in 2022 received a prize for its action plan, part of the Bari Urban Food Policy (BUFP) involving among much else school meals based entirely on Zero Km organic food products, and the creative regeneration of abandoned peri-urban agricultural farmland. It also contained efforts to distribute food to citizens in need through food hubs using the labor of 700 volunteers, projects aiming to eliminate waste by reutilizing food, social agriculture to create work opportunities for unemployed youth using unutilized public property or confiscated mafia property, educational farm projects and university research.[12] These are familiar topics in urban food plans all over the country.

In 2019, the Italian network for local food policies was established, involving researchers, subject experts, public officials, and activists working on relations between food and urban systems. These networks explore urban transformations linked to new food economies and practices, and the governance of local food policies.

Notwithstanding the vibrancy of these many activities, Italian institutional responses to citizen demand for greater food security is fragmented and inconsistent. This trend seems to be representative of the urban food policy efforts in many parts of Europe where local administrations lack the financial and staffing resources required to draw up these plans (Doernberg et al., 2019). In Italy as well the capacity of local government to absorb new policy issues and put them into practice is limited by predetermined institutional conditions which include regulatory support, decision-making mechanisms and available human capacity. Because of these constraints by 2022, only three Italian cities have institutionalized urban food policies (Berti and Rossi, 2022).

Next generation (Piano Nazionale di Ripresa e Resilienza)

The Next Generation fund was approved by the EU in 2020 as an economic recovery package to support member states adversely impacted by the COVID-19 pandemic. In Italy, this was incorporated in the 2021 National Plan for Recovery and Resilience which is supported by a budget of €200b. The green transition, digital transformation, enhancement of citizen health and the struggle against poverty are the main axes of the plan involving a 'new deal' among government, citizens and the productive sector. Solidarity and social justice underpin these development strategies, leaving no one behind, according to the 2030 Agenda for Sustainable Development.

Legambiente, Italy's most important association of environmental activists, recommended priorities for this plan, many of which concern the topics discussed in these pages (Legambiente, 2020). Their first priority is pivoting to renewable energy sources, reducing energy consumption, and closing down fossil fuel burning energy plants. They also recommend important investments in climate change adaptation, supporting sustainable ecological systems through strategies which include urban and peri-urban agriculture, organic and sustainable farming. They advocate the creation of a robust ethical framework militating against the exploitation of labor, and the effective reutilization of confiscated mafia estates to create economic opportunities for citizens pursuing environmentally sound agricultural practices. They note that Italy's organic food system already represent a flourishing sector, involving two million hectares of cultivated land, with 72,000 workers generating annual revenues of €3b. They recommend that organic farming be supported and expanded so it can account for 40% of all cultivated land by 2030. The plan should encourage biodiversity and soil maintenance through traditional and more resilient crops, reducing the use of chemical substances in agriculture, and employing water resources more effectively. These recommendations are consistent with the green revolution which is one of the five missions of the Piano Nazionale di Ripresa e Resilienza (PNRR). Healthcare is another major component of the plan, and Legambiente note the health implications of a diet based on sound agricultural practices, practices which are often organized to involve weak community members, another declared mission of the PNRR.

The question is how much political support will be given to the proposals advocated by Legambiente which concern so many actors in Italian AFN. It is encouraging to see support for robust food policies in many parliamentary initiatives over recent years. A proposal from 2015 to revise the Italian constitution made the effort to reference the right to food, taking as its starting point Art. 25 of the Universal Declaration of Human Rights. Twenty-three nations in the world have included this statement in their constitutions.[13] The Rome Declaration on World Food Security, endorsed at the 1996 World Food Summit organized by the FAO, specifies the right to have secure, sufficient and nutritious food.[14] The proposal for the constitutional reform notes the compelling character of this statement in the current nation with its growing gap between rich and poor. The risk of persisting poverty is said to be one of the highest in Europe. The bill supporting this reform notes that Lombardy Region already passed a provision to guarantee adequate food for all their citizens. The proposal is to expand Art. 32 of the constitution, which guarantees access to health care, to include this statement: 'The Republic safeguards health and food security as fundamental rights for individual and collective interests, guaranteeing free health care and food access to the poor.'

In 2021, the commission of constitutional affairs of the Italian senate revised Art. 9 of the Italian constitution, adding that as well as safeguarding the landscape and the environment, it promotes ecosystem biodiversity in the interests of the nation and its future generations. Already Art. 41 of the Constitution, which concerns economic enterprises, states that they cannot assume forms that cause damage to health or to the environment.[15] All of these policies are consistent with a multifunctional approach to the agri-food system which needs greater sectoral integration to achieve its aims.

Today the collaborative dimension of landscape management is crucial, codesigning and coproducing services, strategies and knowledge. This is becoming a consolidated approach in European urban planning, thanks also to support from Horizon 2020. A plural response to precarity and environmental crisis involves the entire planet in all sustainable urban agendas, which give increasing importance to the social sustainability of behavior and lifestyles (Martinelli et al., 2021). The asymmetries of environmental challenges are increasingly important, as seen in Philip Alston's 2019 concept of 'climate apartheid,' where damaging climate changes have disproportionate impact on the world's poor.[16] A new mutualism of spatial practices must consider urgently the needs of fragile groups and landscapes, reducing inequalities while promoting territorial regeneration, linking together social and environmental justice (De Luca and Cogliati Dezza, 2021). Equitable and sustainable food practices may be the perfect instrument for pursuing these priorities.

Notes

1 https://www.coe.int/en/web/landscape/the-european-landscape-convention (Accessed January 10, 2022).
2 https://www.bizjournals.com/bizjournals/how-to/growth-strategies/2014/06/do-your-corporate-values-pass-the-mafia-test.html (Accessed January 12, 2022).
3 https://foodpolicymilano.org/il-sistema-alimentare/ (Accessed January 10, 2022).
4 https://www.milanurbanfoodpolicypact.org/the-milan-pact/ (Accessed January 14, 2022).
5 http://www.comune.torino.it/amm_com/statuto/statuto.html#art02 (Accessed January 10, 2022).
6 https://www.4cities4dev.eu/eng/adozioni/progetto/3/slow-food-fadiouth-island-salted-millet-couscous-presidium/index.html (Accessed January 10, 2022).
7 https://www.torinovivibile.it/ (Accessed January 12, 2022).

8 https://www.comune.roma.it/web/it/dipartimento-sviluppo-economico-attivita-produttive-progetti.page?contentId=PRG660710 (Accessed July 12, 2022).
9 https://www.gualtierisindaco.it/online/roma-e-tutti-noi/roma-che-vogliamo/#visione-citta (Accessed July 12, 2022).
10 https://www.associazioneterra.it/2021/04/27/food-policy-roma-approvata/ (Accessed July 12, 2022).
11 https://www.cittadelvino.it/articolo.php?id=MTgzMw== (Accessed July 20, 2022).
12 https://www.comune.bari.it/-/alla-citta-di-bari-l-oscar-della-salute-2022-per-la-food-policy-stamattina-il-consigliere-cascella-alla-cerimonia-di-premiazione-ad-ancona (Accessed July 25, 2022).
13 https://www.camera.it/leg17/995?sezione=documenti&tipoDoc=lavori_testo_pdl&idLegislatura=17&codice=17PDL0032630#:~:text=%E2%80%93%20Quello%20dell'accesso%20al%20cibo,famiglia%2C%20con%20particolare%20riguardo%20all (Accessed January 10, 2022).
14 https://www.fao.org/3/w3548e/w3548e00.htm (Accessed August 15, 2022).
15 https://www.ilfattoquotidiano.it/2021/06/07/dal-senato-arrivano-buone-notizie-in-tema-ambientale-anche-se-liter-e-ancora-lungo/6219137/ (Accessed January 10, 2022). https://documenti.camera.it/Leg18/Dossier/Pdf/AC0504.Pdf (Accessed January 10, 2022).
16 https://www.bbc.com/news/world-48755154 (Accessed August 15, 2022).

Bibliography

Alvisi, Mauro (2021) *Concuranza. Trattato generale. Il nuovo paradigma d'intelligenza sociale.* Rome: Mediabooks Editore.

Berti, Giaime and Adanella Rossi (2022) 'Democratic food governance capacity at the local level: the cases of Livorno and Pisa,' *Territory, politics, governance.* https://doi.org/10.1080/21622671.2022.2093265.

Bottiglieri, Maria, Giacomo Pettinati and Alessia Toldo (eds.) (2017) *Turin food policy. Buone pratiche e prospettive.* Turin: Franco Angeli.

Calori, Andrea, Egidio Dansero, Giacomo Pettenati and Alessia Toldo (2017) 'Urban food planning in Italian cities: a comparative analysis of the cases of Milan and Turin,' *Agroecology and sustainable food systems*, 41:8:1026–46.

Calori, Andrea and Francesca Federici (2021) 'Le politiche urbane del cibo come chiave per la sostenibilità,' in Nicola Martinelli, Edoardo Croci and Mariavaleria Mininni (eds.) *Urban@it sesto rapporto sulle città. Le città protagoniste dello sviluppo sostenibile.* Bologna: Il Mulino, pp. 199–209.

Cerreta, Maria and Giuliano Poli (2017) 'Landscape services assessment: a hybrid multi-criteria spatial decision support system (MC-SDSS),' *Sustainability*, 9:8:1311.

Costello, Leesa, Marie-Louise McDermott and Ruth Wallace (2017) 'Netnography: range of practices, misperceptions, and missed opportunities,' *International journal of qualitative methods*, 16:1:1–12.

Council of Europe (2000) *European landscape convention.* Florence: Council of Europe.

De Luca, Sabina and Vittorio Cogliati Dezza (2021) 'La riduzione delle disuguaglianze,' in Nicola Martinelli, Edoardo Croci and Mariavaleria Mininni (eds.) *Urban@it sesto rapporto sulle città. Le città protagoniste dello sviluppo sostenibile*. Bologna: Il Mulino, pp. 223–321.

Dear, Michael J. and Steven Flusty (2000) 'Postmodern urbanism,' *Annals of the Association of American Geographers*, 90:1:50–72.

Di Bella, Elena (2016) 'Food policies at the metropolitan scale,' in Maria Bottiglieri, Giacomo Pettenati and Alessia Toldo (eds.) *Toward the Turin food policy. Good practices and vision*. Milan: Franco Angeli, pp. 23–6.

Di Iacovo, Francesco, Gianluca Brunori and Silvia Innocenti (2013) 'Le strategie urbane: il piano del cibo,' *Agriregionieuropa*, 32. http://agriregionieuropa.univpm.it/it/content/article/31/32/le-strategie-urbane-il-piano-del-cibo (Accessed November 7, 2016).

Doernberg, Alexandra, Paula Horn, Ingo Zasada and Annette Piorr (2019) 'Urban food policies in German city regions: an overview of key players and policy instruments,' *Food policy*, 89:1–13.

FAO (2019) *The Milan urban food policy pact. Monitoring framework*. Rome: Food and Agriculture Organization.

FAO (2022) *Impact of the Ukraine-Russia conflict on global food security and related matters under the mandate of the Food and Agriculture Organization of the United Nations (FAO)*. Rome: Food and Agriculture Organization.

Feagan, Robert (2007) 'The place of food: mapping out the "local" in local food systems,' *Progress in human geography*, 31:1:23–42.

Fondazione Campagna Amica (2020) *La vera agricoltura sociale fa bene all'Italia*. Rome: Fondazione Campagna Amica.

Iovino, Serena (2016) *Ecocriticism and Italy: ecology, resistance, and liberation*. London: Bloomsbury.

Latour, Bruno (2017) *Facing Gaia: eight lectures on the new climatic regime*. Translated by Catherine Porter. Cambridge, UK: Polity Press.

Legambiente (2020) *Recovery plan. Le scelte green indispensabili per un piano italiano capace di futuro*. Rome: Legambiente.

Marino, Davide, Daniele Fattibene, Antonella Marta and Giampiero Mazzocchi (2020) *Città, cibo, sostenibilità. Un tema strategico per l'agenda 2030*. Rome: Alleanza per lo Sviluppo Sostenibile.

Martín, Diego and Rosa de la Fuente (2022) 'Global and local agendas: the Milan Urban Food Policy Pact and innovative sustainable food policies in Euro-Latin American cities,' *Land*, 11:202:1–14.

Martinelli, Nicola, Edoardo Croci and Mariavaleria Mininni (2021) 'Introduzione,' in Nicola Martinelli, Edoardo Croci and Mariavaleria Mininni (eds.) *Urban@it sesto rapporto sulle città. Le città protagoniste dello sviluppo sostenibile*. Bologna: Il Mulino, pp. 15–32.

Morgan, Kevin (2015) 'Nourishing the city: the rise of the urban food question in the Global North,' *Urban studies*, 52:8:1379–94.

Poli, Daniela (2017) 'Food revolution and agro-urban public space in the European bioregional city,' *Agroecology and sustainable food systems*, 41:8:965–87.

Public administration and innovative value chain management 129

Pollan, Michael (2008) *In defense of food: An eater's manifesto.* Westminster: Penguin.

Torino, Comune (2020) *Piano di resilienza climatica.* Turin: Comune di Torino.

Torino, Comune (2020) *Piano strategico infrastruttura verde.* Turin: Comune di Torino.

UN Habitat (2019) *The strategic plan 2020–2023.* New York: UN Habitat.

Vallés-Planells, María, Francisco Galiana and Veerle Van Eetvelde (2014) 'A classification of landscape services to support local landscape planning,' *Ecology and society*, 19:1:44.

Index

Note: Page numbers followed by "n" denote endnotes.

Academy of the Georgofili 10
adverse possession 47, 50
aesthetic dimension 112–13
Agrarian Reform (1950) 70
Agricultural Park, South Milan 5, 116
agri-food market: postwar period 90; vertical coordination 89–90
AGRIFOOD Roma 2030 initiative 122
agri-food system: individual choices and shared outcomes 112; multifunctional approach 126; public policy 112; territorial dimension 115–16
agro-tourism 3, 73
alternative food networks (AFN) 113, 125; globalized agri-food markets 88; social relationships 88; spatial connections 88; success 88
Aniene Regional Park 49
Anti-Camorra Open Day 61
anti-mafia activism 114
antimafia farming: cooperatives 71, 72; farm production 68; mafia property against the mafia 69; non-positivistic approach 67; research questions 67; value chain 66
Antonio Esposito Ferraioli Farm 61
Archdiocese of Monreale 60
Arno river plain 106–8
Artusi, P. 32
ARVAIA 100

Asprinio, grape variety 77
audit culture 6, 7
autochthonous grape varieties 105
autochthonous wheat varieties 105

Baccarini law 30
Bari Urban Food Policy (BUFP) 123
Belice Corleonese 68
Belice Valley 60, 70
Berruti, Gilda 12, 36
Berry, Wendell 112
bioregion 116
blue functions 3
Bolkestein Decree 94
Bologna 78, 98; urban horticulture 55–7
Borghetto San Carlo 81, 82
bottle fermentation 76
Bowles, Camilla 77
Brusca, Giovanni 71
Buitoni 104

Caciofiore di Columella 10
Campagna Amica 97–8
Campania Felix 73
Campania Region 68, 76, 84
Cannaspritz 77–8
care farming 114
Cariplo Foundation 117
car mobility 95
Carosello 95
Casa Chiaravalle 70
Casal di Principe 73

Index

Castellammare di Stabia 103
Cattaneo, Carlo 90
Ceccarelli, Salvatore 106
Centri Sociali Occupati e Autogestiti (CSOA) 79
Cervellati, Pier Luigi 59
Chamber of Commerce 119
Charter for the Italian Network of the Solidary Economy (RES) 99
cheese 7, 10
chemical fertilizers 50–1
cities, as complex ecological systems 43
city of proximity 116
'climate apartheid' 126
Climatic Resilience Plan (*Piano di resilienza climatica*, 2020) 120, 121–2
collaborative approach/dimension 72, 126
Columella, Lucius Junius Moderatus 9, 10
commercial activities 53, 94
commodity outcomes 4, 6, 7, 77
communist party's recreational group (ARCI) 57
community cohesion 84
community gardens 42–4, 50, 122
community services 83
community supported agriculture (CSA) 100
community wellbeing 43
Conad 91, 93, 95, 96
Consolata gardens 47–8
Consortium for Development and Legality 71
Consortium for Legality and Development 72
Consorzi di Bonifica 30
Coop 95–6
Coop Camilla 98, 99
cooperatives 96; antimafia farming 71, 72; competition 114; food 12, 96; social 51, 52, 61, 69, 71, 74, 114
CoopFond 72
coproduction 88, 98
Coraggio 78–83
corporate social responsibility (CSR) 95–6, 114
Cosa Nostra 66

Council of Europe European Landscape Convention (2000) 113
COVID-19 pandemic 4, 12, 44, 78, 124
crop: choice 82; rotation 51

dalla Chiesa, General Carlo Alberto 69
Debord, Guy: *The Society of the Spectacle* 18
De Re Rustica 10
desertification 23, 82
deterritorialization 103–4
Development Education and Awareness Raising (DEAR) program 119
Diana, Don Peppe 74
Diffuse Rural Community (CRD) 101
Diffuse Urban Farm 53
digital communities 113
District of Rural and Social Economy (DESR) 117
Districts for the Solidarity Economy 123
domestic industrial food processors 26
double lowest bid 96–7

Eataly 119
Eatprato 107
ecosystem services 43
edible landscapes, Italy: Ancient Roman principles 20; closed field system 21; commercial sign system 18; EU agricultural planning policy 19; land take 25–9; nut production 31–2; pine nuts 32–3; pine trees 29–31; regional culinary traditions 21–5; regional management plan 26; regional territorial plans 23–4; semi-natural woodlands 21; small-scale natural woodlands 24–5; social scheme 17; three-dimensional mode 18; and visual aesthetics 20–1
Emilia-Romagna 95
environmental damage and responses: citizen association 34; emergency reclamation efforts 33;

flooding forces 36; reafforestation 34; social relationships 34; worst environmental disasters 33–4
environmental stewardship 66
environmental sustainability 112
Epicurean doctrine 1
ethical framework 125
EU directive (2019) 97
Eureka value chain 75–8
European Strategy for Adapting to Climate Change (2013) 121
European Union, agricultural policy 2
Eurospin 97
Experimental Consortium of Wheat Farmers 106
extra virgin olive oil (EVO) 7
extreme micro variation 22

Facciamo un pacco alla camorra 75
Falcone, Giovanni 66
FAO *see* Food and Agriculture Organization (FAO)
farmers markets 97–8
farmers unions 97–8
farming 114; antimafia (*see* antimafia farming); care 114; market-oriented urban farming 53; organic 125; social 12, 66, 68, 69, 83, 114; Urban Farming Global Food Chain 44
farm to fork concept 3
farm-to-fork value generation process 105
Fasci movement 71
filiera 2, 11
fish 99
food: commodity 1; governance 123; grounding of 1; and landscape system 2; logistics 123
food aid 44
Food and Agricultural Organization (FAO) 43, 118
Food Atlas (Atlante del Cibo) 119
food cooperatives 12, 96
food hubs 92, 117
food policy activities 117
Food Smart Cities for Development 119
food sovereignty 101, 120
food waste 117, 118, 123
forest fires 22, 23
Fosso Bravetta Association 48, 49
fragmented landscapes 4–5

Garden Project Team of Venice 55
gastronomic culture 123
gender empowerment 44
Gragnano 9, 102–3
Gran Prato 106–7
Grasseni, Cristina 7
green functions 3
green spaces 25, 46, 47, 51, 59, 79, 83
'Greta Effect' 100
Gualtieri, Roberto 122
Guidelines for the Milan Food Policy (2015–2020) 117

'Health Budget' 74
healthcare 125
History of the Italian Agricultural Landscape (Sereni, E.) 20
holiday farms *(agriturismo)* 105
Horizon 2020 126
horizontal integration 90
horticultural therapy 52
human awareness 6

immigrant vendors 93
Institute for Shelter and Education (IRE) 53
integrated local system 107
internet sales 89
intuition 3
IRE *see* Institute for Shelter and Education (IRE)
Italian Association of Municipalities 121
Italian Communist Party (PCI) 57
Italian community gardens 44
Italian culinary variation 22
Italian Federation of Street Vendors 90
Italian food system 90; procurement system 89
Italian regional diet 99
Italy: first agrarian institute 10; fossil fuels 8; fragmented economic system 114; history 10–11; organized criminality 9;

134   Index

urban-rural linkage 115, 116; wet rice production 23; see also individual entries

Kevin Lynch's image survey methodology 36

labor exploitation 96
Land of Fires 75
landscape 4; management 112, 113, 126; natural 3
La Proposta 51
Latour, Bruno 17
leadership responsibility 72
legality 72
Legambiente 124, 125
Libera Association 69–71, 83–4
'liberated' mafia estates 75
licenses 94
life support services 43
limited-growth model 76
Little Red House 80
local agricultural resources 123
local food policy 118, 124
localized knowledge systems 8
local markets 123
local produce sales 83
Lombardy Region 70

mafia-type criminality 66; Campania 68, 84n5; regions 67; Rome 68; Sicily 68
Mahler, Gustav 6–7
Marino, Ignazio 47, 78, 122
market-oriented urban farming 53
market value 11
metropolitan Naples, urban horticulture 60–2
Michael Porter value chain 2
micro-territorial characteristics 106–7
Milan 94
Milan City Council 118
Milan Food Policy Pact (MUFPP) 115, 116–18
Milan World Expo (2015) 116
Mission of Hope and Charity 60
moral responsibility 113–14
multifunctional agriculture 19
multifunctionality 2, 3, 19, 43, 114

Naples 8, 67, 93
National Agency for the Management and Assignment of Property Seized and Confiscated from Organized Criminality 71
National Plan for Recovery and Resilience (2021) 124
national self-sufficiency 12, 28
National Strategy for Adapting to Climate Change (2015) 121
natural landscape 3
Neapolitan hinterland 61
neighborhood markets *(mercati rionali)* 90–1, 94
neighborhood retail 92–4
neoliberal approach 53
Nestlé 104
netnographic approach 113
New Organized Cooperation (NCO) 73–5
Next Generation fund 124–6
nitrogen fixing plants 51
non-commodity outcomes 6, 7, 77
non-food items 90
nut production 31–2

One Hundred Steps 72
opera a scomputo 85n14
oranges 99
organic farming 125
Orme 58
Orto Capovolto project 59
Orto de' Pecci garden 51
Ortomangione project 101

Pagliaccia, Claudio 13n4
paid employment 73
Palermo, urban horticulture 59–60
Parataxis 18
pasta 8–9; making of 102–3
Pecorino romano 10
pedestrian concept 90
peri-urban: agriculture 91–2, 117–18; horticulture 47; landscapes 4
personal dimension 112
personal gardens 43
personal relationships 90
Petrini, Carlo 98
Piano Nazionale di Ripresa e Resilienza (PNRR) 125

pine nuts 32–3
pine trees 29–31
Pingree's potato patch 44
Pisa 115
Pisapia, Giuliano 117
Placido Rizzotto 70–3
plantations 20–1
policy 2, 5, 6, 12, 13, 19, 21, 57, 60, 78, 112, 113, 115–20
Portella della Ginestra 71
poverty 125
Prefect of Palermo and Libera Association 72
prepared agricultural products (PAPs) 8
private citizen ownership loss 47
procurement 43
protected designation of origin (PDO) 102–3
public food policy: agri-food system 112; Milan Food Policy Pact (MUFPP) 115, 116–18; Turin food plan 115, 118–122
public meetings 117
public space 92
purchasing groups (GAS) networks 99–100

reafforestation 34
reciprocal interdependence, humankind and nature 113
regeneration 30, 33, 34, 37, 58–60, 62, 81, 92, 121, 124, 126
regulation 43
rehabilitation strategy 52
renewable energy sources 125
Resources for Rome 49, 50
retail food procurement 94
Ridolfi, Cosimo 10
Rome 68, 78–80, 84, 92, 94, 122–3; agricultural activities 43, 46; community garden 48; environmental services 46; urban gardens 47; urban horticulture 46–51; Zappata romana 47
Rural City Park 100

Saint Augustine 1
San Giuseppe Jato 71
San Paolo Foundation 119

Saragolla 9, 103, 106
Savorgnano, Michele 53
Senator Cappelli (wheat) 28, 106
school meal plans 123
Science in the Kitchen and the Art of Eating Well (Artusi, P.) 21–2
Second World War 46, 49
Semi di Comunità ('Seeds of the community') 101
Sereni, Emilio 18
sestieri 54
Sevelle, Taja 44
sharing information 90
sheep products 99
short food chain *(filiera corta)* 88
Sicilian Spring 59
Sicily 66, 68, 83, 106
Siena, urban horticulture 51–2
Slow Food 1, 27, 72, 98, 119, 120
small scale retail services 95
Smith, Gregory 12, 85n13, 108n2
social agriculture 114
social cooperatives 51, 52, 61, 69, 71, 74, 114
social ecological system 113
social economy 114
social farming 12, 66, 68, 69, 83, 114
social media 83
The Society of the Spectacle (Debord, G.) 18
solidarity purchasing groups (GAS) 89
squatting property 78
Strategic Plan for Green Infrastructure (*Piano strategico infrastruttura verde,* 2020) 120–1
Strategic Plan of the Metropolitan City of Florence (2030) 108
supermarkets 89, 94–7
supply chain concept 2
sustainable food value chain 2
synergy, humans and environment 3–4

tax incentives 117
Terra Libera 73
territorial systems 88
Third Landscape 5
Together for Aniene 50
Torre Maura 92, 93

tourism 53; agro-tourism 3; ethical 120
trenette 33
Trevi 6
Triticum monococcum 29, 105
Turin Food Plan (TFP) 115, 118–22
'Turin Livable City' 120
Turin, urban horticulture 57–9
Tuscany 10, 52, 104

UAA *see* utilized agricultural area (UAA)
UN 2030 Agenda 119, 124
UN Habitat 112
UPA *see* urban and peri-urban agriculture (UPA)
urban agriculture 11, 117–18, 124
urban and peri-urban agriculture (UPA) 43
Urban Farming Global Food Chain 44
Urban Food Policy (UFP) 115
urban herders 120
urban horticulture 44; background on 43–6; in Bologna 55–7; community support 45; land tenure 45; long-term outlook 45; management 45; in metropolitan Naples 60–2; outcomes 45; in Palermo 59–60; in Rome 46–51; in Siena 51–2; in Turin 57–9; in Venice 52–5; work, equipment 45
urban landscapes 4
urban policies and plans 12

urban–rural continuity 112
urban–rural linkage 115, 116
URP plan 123
utilized agricultural area (UAA) 42

vacant lands 44
Val d'Orcia 105
value chain (VC) 13, 66, 74, 83; analysis 5; approaches 2
value chain experiments 98–102; wheat 102–8
Venice 93; urban horticulture 52–5
Venice World Sustainability Capital 53, 55
vertical coordination 89–90
vertically integrated market 76–7
very high temperature (VHT) techniques 103
villa rustica 20
villa suburbana 20
viticulture 5–6, 10–11

wastelands/wastescapes 34, 44, 61
water bomb 17
wheat 9
white functions 3
wine heritage 119
winemaking facility 72–3, 76, 84
wines 5–6, 10–11
World Food Summit (1996) 125

yellow functions 3

Zappata romana 47
zero-kilometer (0 km) food 98, 101

For Product Safety Concerns and Information please contact our EU
representative GPSR@taylorandfrancis.com
Taylor & Francis Verlag GmbH, Kaufingerstraße 24, 80331 München, Germany

www.ingramcontent.com/pod-product-compliance
Ingram Content Group UK Ltd.
Pitfield, Milton Keynes, MK11 3LW, UK
UKHW021056080625
459435UK00003B/21